This book provides a basic introduction to twenty-five commonly occurring families of flowering plants, chosen for their economic, ornamental and ecological importance. It is designed to enable students of botany and related disciplines to gain some knowledge of the general characteristics of each family and the relationships between them. It will also be helpful to those pursuing courses in botanical illustration, field studies, and other activities requiring a knowledge of flower structure. An introductory section provides basic botanical information which is often assumed to be known and which is essential for a proper consideration of the families themselves. These are described in the second section of the book. For each family, information on its distribution, classification, general features and economic importance precedes a detailed description of a typical representative species. For the larger or more varied families several representative species are included. The text is illustrated throughout with clear and accurate line drawings taken from life and these are accompanied by a written commentary. There are also numerous explanatory diagrams. An exhaustive glossary is provided as an additional aid to the reader.

Common families of flowering plants

Common families of flowering plants

Michael Hickey
and Clive King

CAMBRIDGE
UNIVERSITY PRESS

PUBLISHED BY THE PRESS SYNDICATE OF THE UNIVERSITY OF CAMBRIDGE
The Pitt Building, Trumpington Street, Cambridge CB2 1RP United Kingdom

CAMBRIDGE UNIVERSITY PRESS
The Edinburgh Building, Cambridge CB2 2RU, United Kingdom
40 West 20th Street, New York, NY 10011–4211, USA
10 Stamford Road, Oakleigh, Melbourne 3166, Australia

First published 1997

Printed in the United Kingdom at the University Press, Cambridge

Set in Times

A catalogue record for this book is available from the British Library

Library of Congress Cataloguing in Publication data applied for

ISBN 0 521 57281 9 hardback
ISBN 0 521 57609 1 paperback

Contents

FAMILIES IN SYSTEMATIC ORDER: DICOTYLEDONS 1-19, MONOCOTYLEDONS 20-25

Diagrams

Foreword

The success of 'The 100 Families', as the book by Michael Hickey and Clive King first published in 1981 became popularly known, makes it virtually certain that this new work of theirs will be welcomed by an even wider range of botanical and horticultural users. It is a great pleasure for me to see the book completed and ready for the Press, and to contribute a short Foreword, as I was in at the beginning of the authors' partnership when I was Director of the Cambridge University Botanic Garden.

The decision to make radical changes in the book, and in particular to restrict its coverage to 25 common Angiosperm families, is one which cannot have been taken easily. Some of us would indeed have preferred to have both a third edition of the existing work, updated but not radically revised, as well as this new book concentrating on the 25 most important families and treating them in greater detail. Perhaps - who knows? - we can have both in the future if, as I confidently predict, 'The Common Families' sells really well.

However that may be, we can all unreservedly welcome the new 30-page section on basic botany. It is a fact that there is a large and growing demand from older students, many of them really enthusiastic gardeners and field botanists, who would like to move their hobby from a casual to a more committed interest, and have the enthusiasm and ability to do this. For them this new book caters supremely well. At the same time, it recognises that the university students of today, faced with ever-increasing pressure to learn exciting new areas of botany, may well find a basic, clear, authoritative text covering traditional areas very valuable indeed. I cannot improve, in this Foreword, on my closing remarks in the Foreword to the original '100 Families' written fifteen years ago: 'Generations of students will, I hope, test the value of this book, and come to appreciate its worth.'

S.M. Walters
1 Symonds Lane
Grantchester
Cambridge
CB3 9NU

Acknowledgements

We would like to express our appreciation to Dr S.M. Walters for his advice and encouragement during the writing of this book. We would also like to thank Dr C.D. Pigott, formerly Director of the University Botanic Garden, Cambridge, and members of the staff for their help in providing much of the plant material used for the illustrations. We are indebted to Dr Anne Bebbington of Juniper Hall Field Study Centre, Dorking, for her assistance with the fruit classification chart and general advice; to Dorothy Davies for additional plant material; to the staff of TYPO plus, France Lynch, Stroud, for help with layout, and also to the staff of Cheltenham and Gloucester College of Higher Education. In particular, we would like to thank Robert King for his valuable work with computer equipment which resulted in the production of the camera-ready copy. Finally, we would like to thank Dr Maria Murphy and other members of the Cambridge University Press for their expert advice and assistance in the preparation of this book.

M.H.
C.J.K

Preface

The structure of flowering plants and their grouping into families is a complex subject, but we hope that, by means of a concise text and numerous illustrations, we have provided an interesting and informative introduction to this particular aspect of botany.

The book is divided into two main parts. The first part contains some details of plant structure, nowadays less often taught in schools and colleges than previously, which we feel is nevertheless essential to the understanding of plant families. For this reason, we have included vegetative as well as floral features of flowering plants, information that was standard in botanical text-books of the past. We considered that it would be helpful to the reader to have this basic information readily to hand when dealing with the descriptions of families and their representative plants that follow later and form the major part of the book.

The second part is devoted to descriptions of 25 common plant families, arranged in taxonomic order and chosen for their economic, ornamental and ecological importance. Occasionally, we have made reference to other families in order to clarify a particular point, for example, the often confusing similarity between grasses, rushes and sedges. Each of the texts includes a section on the classification and importance of the family concerned, followed by the detailed description of a typical representative of that family, illustrated by original line drawings showing the parts of the flower, together with their measurements. In the case of the larger or more varied families several examples have been used to show the range of variation, and pages of flowers or fruits added where a family is especially diverse.

The book is intended primarily as a practical manual, and the user is encouraged to have the text to hand as the chosen flower or its alternative is being studied and dissected. Most of the floral parts can be seen either with the naked eye or with a x10 lens, and only rarely will a low-powered microscope be required.

It must be emphasised that the measurements for the floral parts are those of the specimens used for the drawings, and care was taken to ensure that these were typical in form. However, variation within a species or developmental factors may mean that certain parts do not agree exactly with the description or dimensions given.

We hope that this book will be of use to students of botany, both amateur and professional, and also to those pursuing courses in botanical illustration, field studies, and other activities requiring a knowledge of flower structure.

M.H.
C.J.K.

Systematic grouping of plants

The basic unit of classification in the plant kingdom is the species, species being grouped into a genus, genera into a family, families into an order, and so on.

In the list of Families in Systematic Order on page *vii*, the flowering plants are divided according to the usual practice into dicotyledons and monocotyledons. The next division into orders (ending in '-ales') has been omitted since this book covers only 25 families in detail. It will be noticed that all the family names end in '-aceae' with the exception of certain long-established families where we have kept the traditional names. However, the more recently created alternative names for these families, ending in '-aceae', have also been given. In the Classification section of the family texts it will be found that there is quite often, particularly in the case of the larger families, a further division into subfamilies, with names ending in '-oideae', and/or tribes, ending in '-eae'. The endings of these ranks are those prescribed by the International Code of Botanical Nomenclature.

Signs and abbreviations

*	indicates a family not dealt with in detail in this book
∞	indefinite number
×	hybrid
()	united
A	androecium (stamens)
C	corolla
$\overline{\underline{G}}$	gynoecium (ovary inferior)
\underline{G}	gynoecium (ovary superior)
K	calyx
P	perianth
L.S.	longitudinal section
T.S.	transverse section
c.	circa
cm	centimetre(s)
m	metre(s)
mm	millimetre(s)
adj.	adjective
plur.	plural

A measurement given without qualification refers to length. Two measurements connected by 'x' indicate length followed by width.

In the Classification section for each family, the number in brackets after a genus indicates the number of species known throughout the world.

Authorities

At the beginning of each family text the name of the family is followed by a personal name, often in an abbreviated form, e.g. Ranunculaceae Juss. (= Jussieu). Similarly at the beginning of the species text, the name of the species is also followed by a personal name, e.g. *Ranunculus repens* L. (= Linnaeus). This is the method used, in accordance with the International Code of Botanical Nomenclature, to refer to the name of a plant in a more precise way by indicating the person (or 'author') who first validly published that name.

The name of a species will sometimes have a 'double citation', e.g. *Consolida ajacis* (L.) Schur. In this case, the plant was first called *Delphinium ajacis* (by Linnaeus), and later transferred to the genus *Consolida* by Schur. Another use of double citation is when a plant first described as a species is now considered to be of lower rank, e.g. a subspecies or variety. The reverse process, an upgrading in rank, also results in a double citation.

Floral formula

Some of the signs and abbreviations already listed occur in the floral formula, which is a convenient form of 'shorthand' for representing the structure of a flower. The letters K, C, A, and G are used to indicate the whorls of floral parts, beginning with the outermost whorl and working inwards towards the centre of the flower. Where there is no separation into calyx and corolla the letter P is used in place of K and C.

Each letter is followed by one or more figures showing the number of parts comprising each whorl, e.g. A5 indicates that the flower has 5 stamens, while A5 + 5 shows that there are 2 whorls of stamens with 5 in each whorl. Where the number of parts is large and imprecise the sign '∞' is used.

If the number of parts in a whorl is variable, a dash joining 2 figures indicates the range of variation, e.g. A2-4 (in *Arum maculatum*) means that from 2 to 4 stamens may be found in a flower of this species.

The parts forming a whorl are sometimes joined, e.g. G(3) denotes a gynoecium of 3 united carpels.

The position of the ovary is shown by a line above or below the letter G, representing an inferior or superior ovary respectively.

Summary of classification

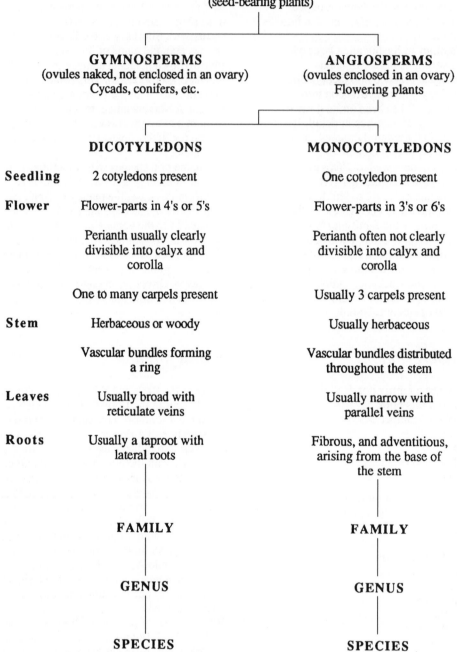

SPERMATOPHYTES
(seed-bearing plants)

GYMNOSPERMS
(ovules naked, not enclosed in an ovary)
Cycads, conifers, etc.

ANGIOSPERMS
(ovules enclosed in an ovary)
Flowering plants

	DICOTYLEDONS	**MONOCOTYLEDONS**
Seedling	2 cotyledons present	One cotyledon present
Flower	Flower-parts in 4's or 5's	Flower-parts in 3's or 6's
	Perianth usually clearly divisible into calyx and corolla	Perianth often not clearly divisible into calyx and corolla
	One to many carpels present	Usually 3 carpels present
Stem	Herbaceous or woody	Usually herbaceous
	Vascular bundles forming a ring	Vascular bundles distributed throughout the stem
Leaves	Usually broad with reticulate veins	Usually narrow with parallel veins
Roots	Usually a taproot with lateral roots	Fibrous, and adventitious, arising from the base of the stem

FAMILY FAMILY

GENUS GENUS

SPECIES SPECIES

N.B. SPECIES may be further divided into SUBSPECIES and VARIETIES.
Horticultural varieties grown in cultivation are known as CULTIVARS, e.g. *Rosa* 'Queen Elizabeth',
Clematis 'Nellie Moser' and *Solanum tuberosum* 'Arran Pilot'.
HYBRIDS sometimes occur between genera, e.g. × *Triticale*, a cross between *Triticum*, Wheat,
and *Secale*, Rye.
The majority occur between species of the same genus. Examples mentioned in this book include
Spiraea × *arguta, Malus* × *domestica* and *Doronicum* × *excelsum*.

General botany

THE PLANT KINGDOM

It is now generally accepted that land-dwelling organisms evolved from unicellular aquatic ancestors such as bacteria and blue-green algae. From these primitive beginnings developed the variety of complex life-forms that are found in the world today. In the course of this period, covering hundreds of millions of years, individual species or groups appeared and disappeared, but the broad categories to which they belonged, once established, have persisted until the present time. Traditionally all forms of life were considered to be either animals or plants, but more recent research has shown that this simple classification is no longer satisfactory. Indeed there are strong grounds for excluding viruses, bacteria, blue-green algae, and even fungi from the plant kingdom. This new, more restricted concept of the plant kingdom includes the following groups:

1. Non-vascular plants
 (i) algae (unicellular or multicellular freshwater or marine plants, e.g. seaweeds)
 (ii) bryophytes (mosses, liverworts)

2. Vascular plants (plants having a vascular system for conducting water and food solutions)
 (i) pteridophytes (ferns, horsetails, clubmosses)
 (ii) gymnosperms (conifers, cycads)
 (iii) angiosperms (flowering plants)

As may be seen from the geological time scale, non-vascular plants were the first to appear, followed at intervals by various kinds of vascular plants. The earliest of these were the pteridophytes, spore-bearing plants with a preference for damp and shady habitats. Then came the gymnosperms, woody plants bearing cones with naked seeds (i.e., not enclosed in an ovary). In this group pollen-grains from the male cones are carried by the wind directly on to the ovules (unfertilised seeds) produced on the scales of the female cones. The most recent group to arise are the angiosperms or flowering plants which have their ovules enclosed in a protective structure, the ovary. This is usually extended upwards to form the style and stigma. Pollen-grains landing on the stigma can only reach the ovules by forming a tube which grows down through the style and into the ovary.

MILLIONS OF YEARS AGO	GEOLOGICAL PERIOD	NON-VASCULAR PLANTS		VASCULAR PLANTS		
		Algae	Bryophytes	Pteridophytes	Gymnosperms	Angiosperms
			Mosses Liverworts	Horsetails, Clubmosses, Ferns	Conifers, Cycads	Flowering Plants
2.5	Quaternary					
65	Tertiary					
135	Cretaceous					
190	Jurassic					
225	Triassic					
280	Permian					
345	Carboniferous					
395	Devonian					
430	Silurian					
500	Ordovician					
570	Cambrian					

Fig. 1. Geological time-scale.

1

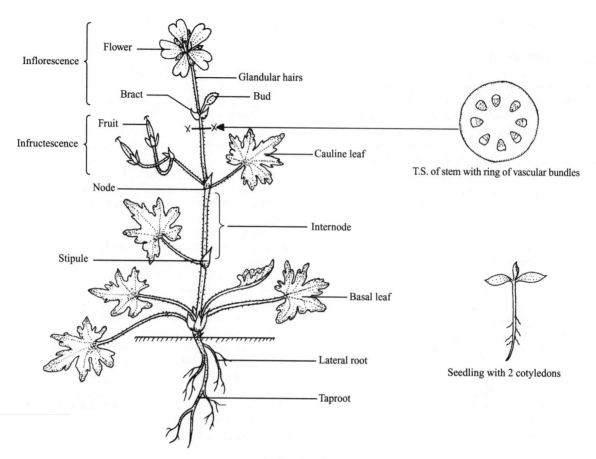

T.S. of stem with ring of vascular bundles

Seedling with 2 cotyledons

Fig. 2. Main parts of a dicotyledon: *Geranium* (Geraniaceae).

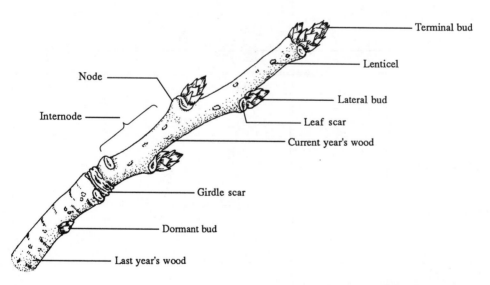

Fig. 3. Stem of a woody dicotyledon: *Prunus* (Rosaceae).

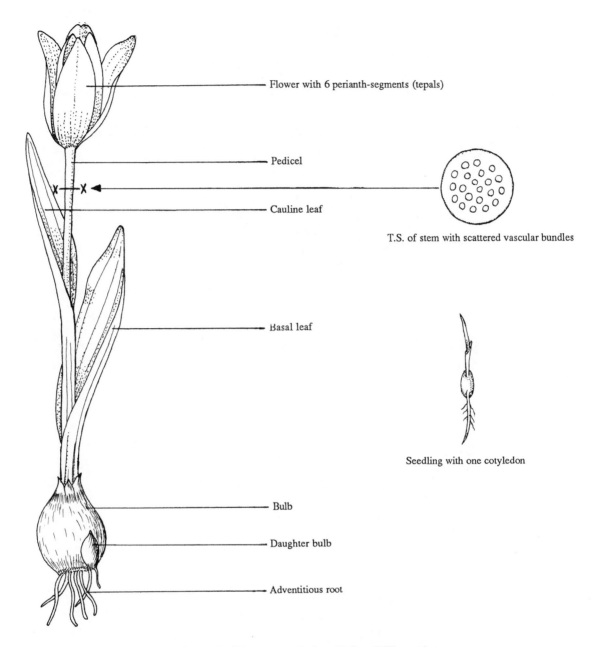

Flower with 6 perianth-segments (tepals)

Pedicel

Cauline leaf

T.S. of stem with scattered vascular bundles

Basal leaf

Seedling with one cotyledon

Bulb

Daughter bulb

Adventitious root

Fig. 4. Main parts of a monocotyledon: *Tulipa* (Liliaceae).

L.S. of *Geranium*, a flower with 5 free sepals, 5 free petals, and a superior ovary.

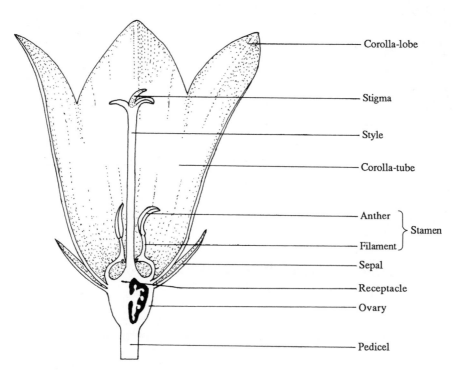

L.S. of *Campanula*, a flower with 5 united sepals, 5 united petals, and an inferior ovary.

Fig. 5. Flower of a dicotyledon (parts usually in multiples of 4 or 5 or else numerous).

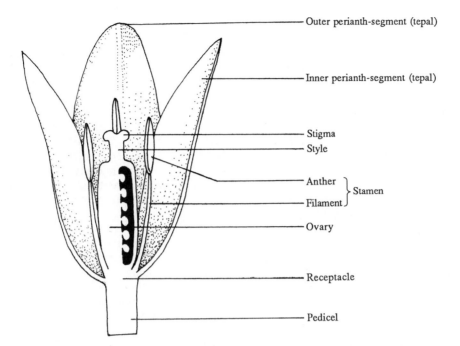

Outer perianth-segment (tepal)

Inner perianth-segment (tepal)

Stigma
Style

Anther ⎫
 ⎬ Stamen
Filament ⎭

Ovary

Receptacle

Pedicel

L.S. of *Tulipa*, a flower with a superior ovary.

Outer perianth-segment (tepal)

Inner perianth-segment (tepal)

Stigma
Style
Anther ⎫
 ⎬ Stamen
Filament ⎭

Receptacle

Ovary

Pedicel

L.S. of *Leucojum*, a flower with an inferior ovary.

Fig. 6. Flower of a monocotyledon (parts usually in multiples of 3).

Note: In many monocotyledons, the perianth is not easily divisible into calyx and corolla, and the term 'perianth-segments' or 'tepals' is used to include both sepals and petals.

THE FLOWERING PLANTS

Classification and evolution

The classification of the flowering plants or angiosperms is based upon comparative study of the structure of their flowers and fruits. Thus, one of the most obvious differences between plants, namely whether they are trees, shrubs, or herbs, is not a character used in the main classification into different families, and indeed many important families, such as Rosaceae and Compositae, contain both woody and herbaceous representatives. But it is significant, and also useful in identification, that certain families, such as those comprising the catkin-bearing group or Amentiferae, are wholly or very largely composed of woody plants, whilst others are entirely herbaceous. It is generally assumed that the change from woody to herbaceous forms, and vice versa, has occurred many times within different evolutionary lines of flowering plants. Many botanists also hold the view that the primitive angiosperm was a dicotyledon with woody stems and large, terminal flowers like the modern *Magnolia*, but this theory has not been proved to the satisfaction of all. Neither is it certain how the monocotyledons originated, though it may be that the common ancestor of both monocotyledons and dicotyledons was some shrubby angiosperm with primitive floral structure belonging to a group that is now completely extinct. Although the form of the earliest flowering plants is still unknown, it seems clear from the fossil record that they arose in the early Cretaceous period (120 million years ago) and that before the end of that period (80-90 million years ago) they had ousted the conifers and cycads from their position of dominance and established themselves as the characteristic form of land plant, a place they have continued to hold to the present day.

Structure of the whole plant

With very few exceptions, growth in flowering plants is maintained underground by a **root** system and above ground by a **shoot** system. Root systems are divided into two main types. The first type, typical of dicotyledons, consists of a vertical **taproot** that has developed from the radicle of a seedling, and **lateral roots** that have grown out from the central part of that root. The second type, common in monocotyledons, occurs when the primary root system is replaced by fibrous **adventitious roots**. It is also found in species of dicotyledons where stolons or runners root at the nodes, e.g. *Fragaria* × *ananassa* (Strawberry), and in climbing plants such as *Hedera helix* (Common Ivy) where the climbing stems produce adventitious roots in order to cling to supports of various kinds.

Roots, and other parts of plants too, sometimes act as **storage organs**, enlarging as they do so. A particular structure may appear more frequently in one family or genus than in another. For example, root tubers are a feature of the genus *Ophrys* (Orchidaceae), and rhizomes are common in many species of *Iris*, while bulbs are common to all lilies. These organs, as well as being able to store food during unfavourable climatic periods, are in many cases very efficient methods of vegetative reproduction. Some of the types of root and storage organ most often found are listed below:

Taproot. The main root becomes swollen to act as an anchorage and to store food. From this primary root small lateral roots grow out to take in water and nutrients from the soil, e.g. *Pastinaca sativa* (Parsnip) and *Taraxacum officinale* (Dandelion).

Tuberous roots. Root branches develop near the base of the plant and become swollen. These root tubers are often modified adventitious roots and may be egg-shaped or cylindrical in shape. They occur in both dicotyledons, e.g. *Dahlia*, and in monocotyledons, e.g. *Ophrys* (Bee and Fly Orchids).

Buttress, stilt, or prop roots are found mainly in the tropical flora, e.g. mangroves. An example from cooler regions is the prop roots that grow out from the 2 or 3 nodes above the surface of the ground to support the tall stem of *Zea mays* (Maize).

Epiphytic roots are aerial roots produced by plants growing on the branches of trees. Many tropical orchids and ferns are epiphytic and their roots obtain water and nutrients from the moisture of the atmosphere.

Stem tubers are swollen underground stems. They differ from root tubers, for when exposed to light they may produce chlorophyll, as when potatoes turn green.

Swollen stems. The genus *Brassica* (Cruciferae) provides several examples. In Kohlrabi, a variety of *B. oleracea*, the base of the stem is swollen, and, together with the attached leaves, forms a useful food crop. In the Turnip (*B. rapa*), the swollen portion is mainly the transition area between the root and the stem, known as the **hypocotyl**. In the Swede, (*B. napus* var. *napobrassica*), both the hypocotyl and the base

Fibrous roots of a Chickweed seedling,
Cerastium sp. (Caryophyllaceae)

Adventitious roots of a grass
(Gramineae)

Taproot of a Parsnip,
Pastinaca sativa (Umbelliferae)

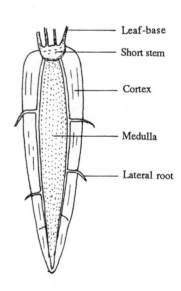

Root tubers of Lesser Celandine,
Ranunculus ficaria (Ranunculaceae)

Root tubers of Bee Orchid,
Ophrys apifera (Orchidaceae)

L.S. of a taproot

Fig. 7. Roots and root tubers.

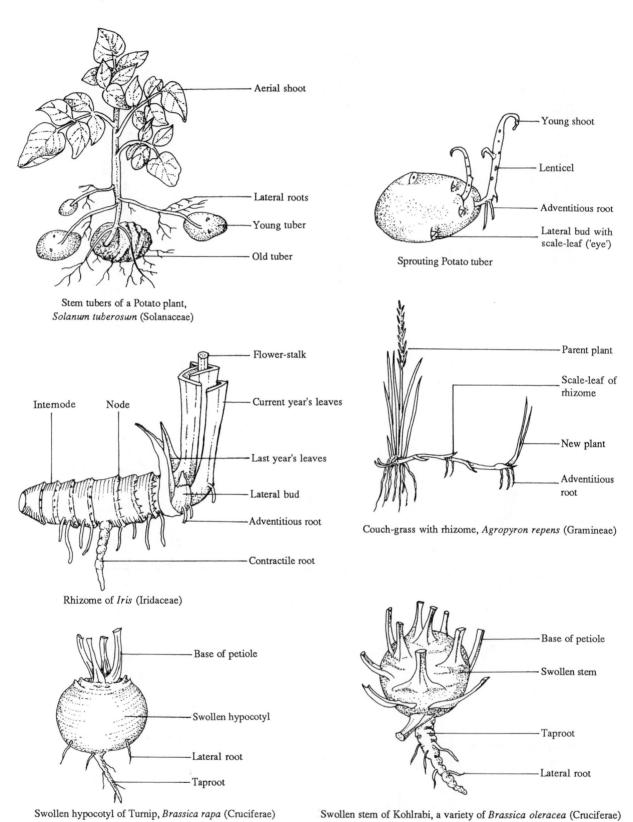

Aerial shoot

Lateral roots

Young tuber

Old tuber

Stem tubers of a Potato plant,
Solanum tuberosum (Solanaceae)

Young shoot

Lenticel

Adventitious root

Lateral bud with
scale-leaf ('eye')

Sprouting Potato tuber

Flower-stalk

Current year's leaves

Last year's leaves

Lateral bud

Adventitious root

Contractile root

Internode Node

Rhizome of *Iris* (Iridaceae)

Parent plant

Scale-leaf of
rhizome

New plant

Adventitious
root

Couch-grass with rhizome, *Agropyron repens* (Gramineae)

Base of petiole

Swollen hypocotyl

Lateral root

Taproot

Swollen hypocotyl of Turnip, *Brassica rapa* (Cruciferae)

Base of petiole

Swollen stem

Taproot

Lateral root

Swollen stem of Kohlrabi, a variety of *Brassica oleracea* (Cruciferae)

Fig. 8. Stem tubers and rhizomes.

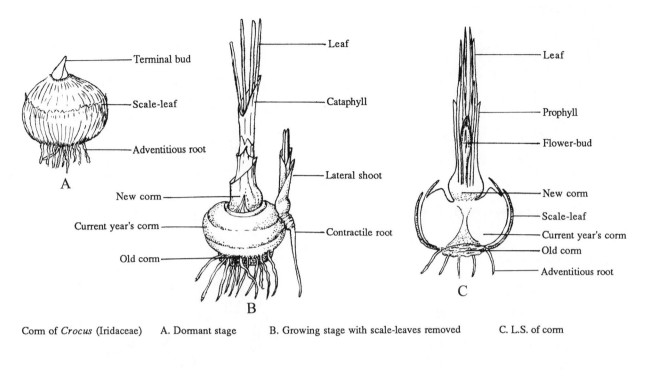

Corm of *Crocus* (Iridaceae) A. Dormant stage B. Growing stage with scale-leaves removed C. L.S. of corm

Bulb of Lily, *Lilium* (Liliaceae)

Inflorescence of Garlic with bulbils *Allium sativum* (Liliaceae)

Swollen leaf-bases of Florentine Fennel,
Foeniculum vulgare var. *dulce* (Umbelliferae)

L.S. of Onion, *Allium cepa* (Liliaceae)

Fig. 9. Corms and bulbs.

Dichasial cymes

Simple cyme *Cerastium* (Caryophyllaceae)

Compound cyme *Stellaria* (Caryophyllaceae)

Monochasial cymes

Simple cyme *Geum* (Rosaceae)

Spiralled cyme *Myosotis* (Boraginaceae)

Racemose inflorescences

Spike *Cephalanthera* (Orchidaceae)

Raceme *Digitalis* (Scrophulariaceae)

Fig. 10. Inflorescences (1).

Racemose inflorescences

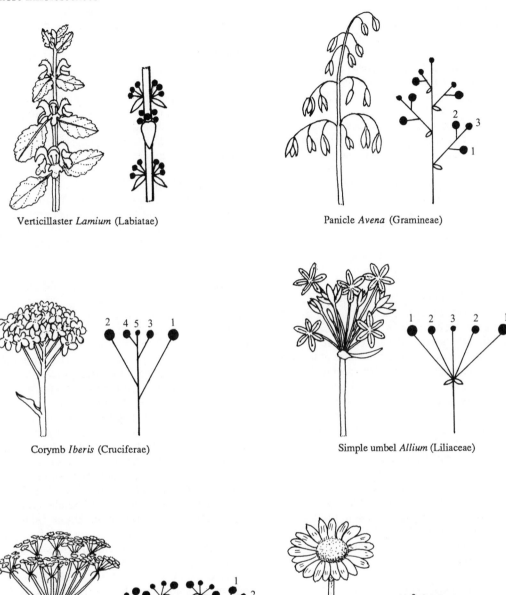

Verticillaster *Lamium* (Labiatae)

Panicle *Avena* (Gramineae)

Corymb *Iberis* (Cruciferae)

Simple umbel *Allium* (Liliaceae)

Compound umbel *Heracleum* (Umbelliferae)

Capitulum *Doronicum* (Compositae)

Fig. 11. Inflorescences (2).

of the stem are swollen.

Rhizomes are creeping stems that spread horizontally, usually underground, as in *Agropyron repens* (Couch-grass), but occasionally on the surface, as in many species of *Iris*. *Iris* rhizomes, as well as *Crocus* corms and lily bulbs, have **contractile roots** as well as adventitious roots to ensure efficient anchorage.

Corms are short, non-creeping, swollen stems that renew themselves every year, e.g. *Crocus* and *Gladiolus*. These corms are surrounded by scale leaves, and the new corm is formed above the old corm. In *Crocosmia* (Montbretia), corms are formed at the tip of long rhizomes. The *Cyclamen* 'corm' differs in being naked, i.e. not enveloped in scale leaves, and in lasting for several years. It is in fact a tuber that increases in size with age and, in some species, may then divide and establish daughter plants.

Bulbs can be divided into 2 categories, those which have an outer coat or 'tunic' of membranous leaves, e.g. *Allium*, *Narcissus*, and *Tulipa*, and those where the fleshy scales are naked, as in species of *Lilium*. At the base of the bulb is a much reduced stem or 'disc' from which grow numerous adventitious roots. In some plants, bulbs may be formed in parts growing above the ground, either in the inflorescence, as in certain species of *Allium*, or in the leaf-axils, as in some species of *Lilium*. Small bulbs formed in this way are known as **bulbils**.

Swollen leaf-bases. In *Foeniculum vulgare* var. *dulce* (Florentine Fennel) the leaf-stalks widen out at the base and envelop each other to form a bulb-like structure. However, most of the food storage takes place in the taproot from which the swollen leaf-bases arise.

The shoot system consists of one or more stems bearing leaves that are arranged spirally or in pairs, or more rarely in whorls of three or more, along each stem. The point from which a leaf arises is called a **node**, and the portion of the stem between two nodes is described as an **internode**. The angle formed by a leaf and its parent stem is known as an **axil**, and buds appearing in this position are called **axillary**. A bud may also be found at the apex of the stem, in which case it is described as **terminal**. Each bud is, in fact, a shoot with very short internodes and immature leaves. Some buds do not develop further but, if they do so, the internodes lengthen and the leaves enlarge, forming a typical shoot. A flower may be regarded as a bud which is

modified for the purpose of reproduction of the parent plant. The outermost leaves have changed little, retaining their leaf-like appearance and protective function. The inner leaves, on the other hand, have in many cases undergone considerable modification in shape and colour, and have become highly specialised in order to perform a variety of functions necessary for the successful propagation of the plant concerned. These modified leaves are now referred to as **floral parts**. In the more primitive families such as *Magnoliaceae, the floral parts are many or **indefinite** in number and are arranged on an elongated axis. In more advanced families there is a reduction in the number of parts which then arise from a much shortened axis. In either case, the production of a flower means that growth of the stem concerned is now limited, although it may continue to elongate sufficiently in order to allow the flower-cluster or **inflorescence** to develop fully.

Growth of a stem by apical extension is termed **monopodial** and results in an inflorescence known as a **raceme**. In contrast to this, the main stem may cease to lengthen owing to a flower being formed at its apex. If this occurs, side or lateral branches may arise from buds below this flower. This is called **sympodial** growth and results in an inflorescence known as a **cyme**. Determining whether an inflorescence is racemose or cymose may sometimes be difficult, but they are usually distinguishable by the order of flower development. In a raceme the youngest flower is situated at the apex of the stem, while in a cyme the oldest flower occupies this position. Grouping the flowers together into an inflorescence, of whatever kind, renders them more conspicuous and is an important aid to pollination. The term inflorescence is often used solely for a cluster of individual flowers, but, correctly employed, it includes the stem from which the flowers arise which is termed the **peduncle**. Sometimes, however, the flowers are solitary on a stem as in *Galanthus* (Snowdrop). If an individual flower has no stalk it is said to be **sessile**, but if a stalk is present this is known as a **pedicel**. There may often be scale-like or leaf-like structures arising from the peduncle or pedicel. These are termed **bracts** or **bracteoles**, and are usually simple in shape compared with the foliage leaves. Where bracteoles are present in dicotyledons, there are usually two situated opposite each other on the pedicel, but only a single bracteole may be present in monocotyledons. In many inflorescences each pedicel arises from the axil of a bract on the peduncle. In the Compositae, and also the *Dipsacaceae (Scabious family), the

inflorescence is condensed to a head of sessile flowers known as a **capitulum** which is borne on the thickened and flattened apex of the peduncle. The bracts of these flowers are crowded into one or more whorls round the capitulum, and together form an **involucre**.

Structure of the flower and fruit

The flower is a specialised structure evolved for seed formation, a process that is normally the result of a sexual union. Some plants, however, have the ability to form seeds without the aid of fertilisation. This form of reproduction, which is found in several common genera of Rosaceae (e.g. *Alchemilla*, *Rosa*, and *Rubus*) and Compositae (e.g. *Hieracium* and *Taraxacum*) is known as **apomixis** (more strictly **agamospermy**). The floral axis on which the flower-parts are arranged is termed the **receptacle** or **torus**. Where a single flower is borne on a pedicel, the receptacle is the upper end of the flower-stalk. In the Compositae and *Dipsacaceae the flowers have no individual stalks but are arranged in a head at the top of the peduncle. Here it is the upper end of the peduncle that is called the receptacle. In either case, the receptacle is often enlarged and may be flat, concave or convex in shape. As a rule, the individual flower consists of male and female reproductive organs surrounded by a **perianth** that is usually divisible into a **calyx** and a **corolla**. If only the male reproductive organs are present the flower is described as **staminate**, and if only the female organs are present it is termed **pistillate**. In some flowers the calyx or the corolla may be absent, but in the majority of species all four kinds of floral parts are present. Working from the outside towards the centre of the flower, the floral parts are as follows:

THE PERIANTH

1. *CALYX*. The calyx forms the outer whorl of the perianth and the separate parts are called **sepals**. These are usually green and leaf-like, and they enclose and protect the inner floral parts in the bud stage. If the sepals are joined together, the united portion is called the **calyx-tube** and the free portions the **calyx-lobes**. Sometimes the sepals are coloured and may then, like the petals, attract insects. Such sepals are described as **petaloid**. In some families, e.g. Compositae, the calyx has been modified into a group of hairs and is called a **pappus**.

2. *COROLLA*. The corolla forms the inner whorl of the perianth and the separate parts are called **petals**. These are usually coloured and serve to attract insects or, in some exotic species, birds, for the purpose of pollination. If the petals are joined together, the united portion is called the **corolla-tube** and the free portions the **corolla-lobes**. Sometimes the petals are green and they are then described as **sepaloid**. In some cases both sepals and petals are similar in shape and colour. These perianth-segments are often known as **tepals**.

THE REPRODUCTIVE ORGANS

1. *ANDROECIUM*

Structure. The **androecium** is the collective term for the male organs of the flower, and the individual parts are called **stamens** if functional and **staminodes** if non-functional. Each stamen has two main parts: the **anther**, and its supporting stalk, the **filament**. The anther is usually composed of two lobes or **thecae**, each comprising a pair of pollen-sacs or **loculi** which contain the pollen. The lobes are joined together by a continuation of the filament known as the **connective**. Sometimes, as in the family Malvaceae, the anther has only one lobe. It is then termed **monothecous**, in contrast to the normal, **dithecous** condition.

Stamen attachment. The stamens may be attached to the receptacle or to the petals. In the latter case they are called **epipetalous**. Where the stamens are in two whorls, the inner whorl opposite the petals and the outer whorl opposite the sepals, the flower is termed **diplostemonous**. In some families, e.g. Geraniaceae, the position of the whorls is reversed, this arrangement being described as **obdiplostemonous**.

Anther attachment.
(i) Fixed anthers. The base of the anther may be fixed to the apex of the filament, i.e. **basifixed**, or the filament may be attached to the back of the anther, i.e. **dorsifixed**. Sometimes a considerable portion of the anther is united with the filament and it is then incapable of separate movement.
(ii) Moveable anthers. If a dorsifixed anther is attached only to the apex of the filament, it is able to move independently and is termed **versatile**. In certain flowers the filaments are so slender that they bend over at the apex and the anthers are then described as **pendulous**. Versatile and pendulous anthers are frequently found in wind-pollinated flowers.

Dehiscence. The opening or dehiscence of the anthers is usually lengthwise or **longitudinal**. In this case the opening occurs between each pair of pollen-sacs, and pollen from both sacs of an

Petals free (polypetalous)

Regular (Cruciferae) Irregular (Leguminosae) Regular (Liliaceae) Irregular (Orchidaceae)

Petals united (gamopetalous)

Regular, campanulate (Campanulaceae) Regular, funnel-shaped (*Convolvulaceae) Regular, rotate (Solanaceae) Regular, tubular (Boraginaceae) Regular, urceolate (Ericaceae)

Regular, salver-shaped (*Polemoniaceae) Irregular, personate (Scrophulariaceae) Irregular, labiate (Labiatae) Irregular, ligulate (Compositae)

Fig. 12. Corollas.

Open Valvate Imbricate Convolute

Contorted Quincuncial Imbricate-descending (vexillary) Imbricate-ascending

Fig. 13. Aestivation.

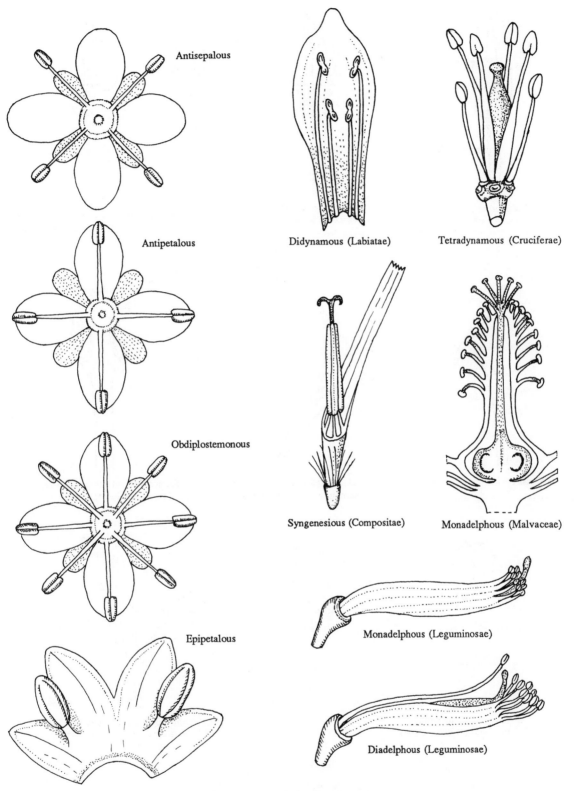

Antisepalous

Antipetalous

Obdiplostemonous

Epipetalous

Didynamous (Labiatae)

Tetradynamous (Cruciferae)

Syngenesious (Compositae)

Monadelphous (Malvaceae)

Monadelphous (Leguminosae)

Diadelphous (Leguminosae)

Fig. 14. Stamen arrangement.

Basifixed anther
(dorsal view)

Basifixed anther with
dehiscence lengthwise
(ventral view)

A B
Basifixed anthers with poricidal dehiscence, *Erica*
(Ericaceae) A. *E. vagans* B. *E. tetralix* (with awns)

Dorsifixed anther (non-versatile)

Dorsifixed anther (versatile)
common in grasses (Gramineae)

Anther with valvate dehiscence,
Laurus nobilis (*Lauraceae)

2 views of connate anthers
of *Saintpaulia* (*Gesneriaceae)

Anther with nectar-spur
Viola (*Violaceae)

Several anthers on a much-branched filament
of *Ricinus communis* (Euphorbiaceae)

Fig. 15. Anthers.

Introrse Extrorse

Fig. 16. Dehiscence of stamens.

Ovary superior, flower hypogynous Ovary superior, flower perigynous Ovary inferior, flower epigynous

Fig. 17. Ovary position.

Free (apocarpous)

Several free carpels, One free carpel, Several free carpels,
Aconitum (Ranunculaceae) *Lathyrus* (Leguminosae) *Geum* (Rosaceae)

United (syncarpous)

Styles free, *Cerastium* Single style with branches, Single style with capitate Style absent and stigma
(Caryophyllaceae) *Juncus* (*Cyperaceae) stigma, *Primula* (Primulaceae) sessile, *Arabis* (Cruciferae)

Fig. 18. Carpels.

anther-lobe is released from the same split. Another form of dehiscence (**poricidal**) is exhibited by some members of the Ericaceae, pollen being shed through terminal pores. Yet another kind of pollen-shedding, **valvate** dehiscence, is shown by the *Lauraceae (Laurel family). In this family a flap-like valve opens upwards to liberate the pollen. Dehiscence of an anther is described as **introrse** if it is towards the centre of the flower, and **extrorse** if it is away from the centre. In some families the anther opens at the side, and this is known as **lateral** dehiscence.

Fusion. The stamens may be free from each other or they may be joined together in some way. In the Malvaceae and in some of the Leguminosae all the filaments are united for most of their length, forming a tube round the gynoecium. The stamens are then termed **monadelphous**. In other members of the Leguminosae the stamens are united by their filaments into 2 groups and are called **diadelphous**. Yet another arrangement is found in the *Guttiferae (St John's-wort family). Here, in the genus *Hypericum*, the stamens are grouped into 3 or 5 bundles or **fascicles**, the filaments in each group being joined together only at their base. By contrast, members of the Compositae have the filaments free but the 5 anthers are united. This condition is described as **syngenesious**.

Relative length of stamens. In most families the stamens of a particular flower are all about the same length, but in the Labiatae **didynamous** stamens (2 long and 2 short) occur frequently, and in the Cruciferae **tetradynamous** stamens (4 long and 2 short) are the typical condition.

2. GYNOECIUM
Structure. The **gynoecium** is the collective term for the female organs of the flower, and the individual parts are called **carpels**. In some species, e.g. *Vicia faba* (Broad Bean), the gynoecium consists only of a single carpel. In most plants, however, the gynoecium is composed of several carpels. It is termed **apocarpous** if the carpels are free, as in *Ranunculus repens* (Creeping Buttercup), and **syncarpous** if they are united, as in *Solanum dulcamara* (Woody Nightshade). Each carpel consists typically of an **ovary**, a **style**, and a **stigma**. In cases where the gynoecium is syncarpous, the seed-bearing parts are collectively called the ovary.

The ovary. In an ovary formed from one carpel, e.g. Leguminosae, there is a central cavity, the **loculus** or **cell**, which contains the unfertilised seeds or **ovules**. Each of these is attached by a stalk, the **funicle**, to the **placenta**, a specialised region of the ovary-wall. In a syncarpous ovary, the number of carpels involved in its formation may often be determined by making a transverse section of the ovary and then counting the number of loculi present. Occasionally, the syncarpous ovary is unilocular, and then the number of carpels may be indicated by the number of styles, as in some members of the *Passifloraceae (Passion-flower family). If, however, the carpels are entirely fused into one ovary, one style, and one stigma, their number may yet be ascertained by counting the lines or **sutures** on the outside of the ovary where the carpels are coalescent. The number of valves into which the mature fruit splits may also reveal the number of carpels concerned. The ovary is described as **superior** or **inferior** according to its position on the receptacle in relation to the other floral parts.

The ovule and its orientation. The ovule is made up of central tissue known as the **nucellus** in which the embryo sac develops and, surrounding the nucellus, the **integuments** which form the outer covering of the ovule. There is a minute opening in the integuments known as the **micropyle** which, by exposing a portion of the nucellus, contributes to the process of fertilisation. In addition, at the seed stage, it permits the passage of water essential to germination. The ovule is positioned in a particular way, and may be **orthotropous**, **anatropous**, **campylotropous**, or **amphitropous**. The above must not be confused with what is called **placentation**, the arrangement of ovules within the ovary.

The style and stigma. The style usually arises from the apex of the ovary, but in some families, e.g. Labiatae, it may arise from the base, in which case it is termed **gynobasic**. At or near the top of the style is an area receptive to pollen, the **stigma**. In some members of the *Papaveraceae (Poppy family) the styles are absent, and the stigmas are situated directly on the ovary, appearing as a series of radiating lines. Where the carpels are free there is usually a single style present on each carpel, and in flowers which have their carpels united only at the base, the styles remain more or less free. But where the carpels are completely united, their styles are often united also, forming a single style with one fused or several free stigmas.

Transverse sections

Chalaza

Nucellus
with embryo

Integuments

Funicle

Micropyle

Anatropous

Free-central

Parietal

Orthotropous

Campylotropous

Axile

Marginal

Longitudinal sections

Amphitropous

Apical

Basal

Fig. 19. Ovules.

Fig. 20. Placentation.

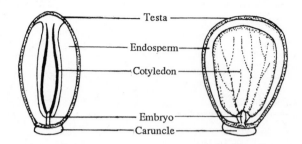

L.S. of seed of the dicotyledon *Ricinus communis*, Castor-oil Plant, cut in 2 ways.

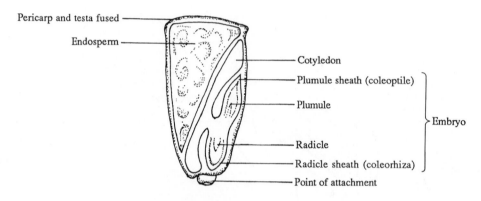

L.S. of seed of the monocotyledon *Zea mays*, Maize.

Fig. 21. Endospermic seeds.

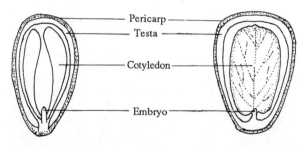

Fig. 22. L.S. of fruit of *Helianthus annuus*, Sunflower, cut
in 2 ways to show the testa of the seed joined to the pericarp.

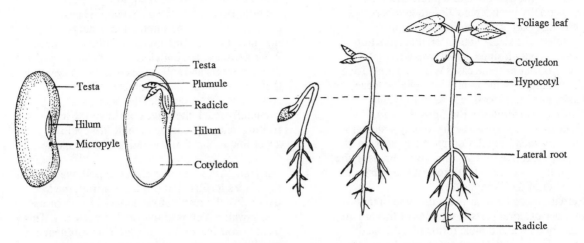

Epigeal germination and seed structure of *Phaseolus vulgaris*, French or Kidney Bean.

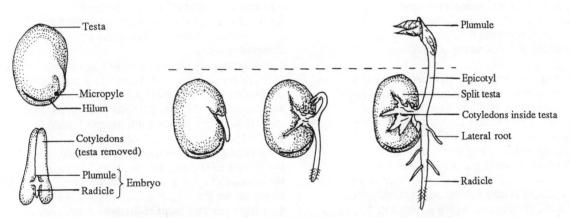

Hypogeal germination and seed structure of *Vicia faba*, Broad Bean.

Fig. 23. Germination of dicotyledons.

Hypogeal germination of *Zea mays*, Maize.

Fig. 24. Germination of monocotyledons.

Variation in the number of floral parts. In the course of the evolution of a plant, there has occasionally been a reduction in numbers of a particular part, sometimes resulting in its complete absence. Quite often this reduction involves only a single structure, like petals or stamens, but in some cases it is more extensive. For example, in the family Scrophulariaceae, the basic number of parts is 5, as in the genus *Verbascum* (Mullein). But *Scrophularia* (Figwort) and *Penstemon* have only 4 stamens, and *Veronica* only 2. Complete suppression of the reproductive organs can also occur, as in the sometimes sterile outer florets of *Centaurea*, a member of the Compositae.

Perhaps more familiar is the increase in the number of floral parts. Examples of this are the semi-double and double forms of cultivated roses, fuchsias and pelargoniums, where stamens or other floral parts have become modified to form petals. This situation may arise through natural mutation, but increasingly in modern times it has occurred through deliberate hybridisation for commercial purposes.

The seed. In every angiosperm seed there is an amount of nutritive tissue called the **endosperm** that is absorbed into the embryo as the seed develops. This process takes place at varying rates, seeds being classified into 2 groups according to the rate of absorption. In **endospermic** seeds, e.g. *Ricinus communis* (Castor-oil Plant) and *Zea mays* (Maize), part of the endosperm lasts at least until germination. But if the process is completed more rapidly, so that the endosperm has disappeared by the time of germination, the seed is described as **non-endospermic**. Examples of this type are *Vicia faba* (Broad Bean) and *Phaseolus vulgaris* (French Bean).

It should be noted that the so-called 'seeds' of the Compositae and the Gramineae are in fact fruits, since the fruit-wall (pericarp) is fused to the seed-coat (testa).

Germination. The conditions required for the successful germination of seeds differ considerably but, in general, a period of dormancy, an adequate supply of water and oxygen, and an optimum temperature are needed in order to initiate the process. Two kinds of germination are recognised, but in both cases the need is for a photosynthetic surface to maintain the growth of the **radicle** (root). In **epigeal germination**, the cotyledons or seed-leaves emerge above the ground and turn green, and the **plumule** (shoot) is often only rudimentary. A notable exception to this is *Phaseolus vulgaris*

where the plumule develops and becomes quite conspicuous within the seed. In **hypogeal germination**, the cotyledons remain below ground, and it is the plumule that emerges above the surface to become green and photosynthetic. After germination, both kinds of seedlings will develop stems and foliage leaves, thus increasing their potential for photosynthesis and growth. If conditions are favourable, the plants will eventually reach the stages of flowering and fruiting, and will complete their life-cycle according to the characteristics of their species.

The fruit. After a flower has been pollinated and the ovules fertilised, it produces a fruit, the true fruit being the result of secondary growth of the ovary-wall which is known as the **pericarp**. This usually consists of 3 layers, an inner **endocarp**, a middle **mesocarp**, and an outer **epicarp** or **exocarp**. Enclosed within the fruit are the seeds which have developed from one or more fertilised ovules. In some plants fruits are formed in the absence of fertilisation, a condition known as **parthenocarpy**.

Fruits are usually classified according to whether they are dry or fleshy at maturity. Dry fruits can be divided into those that split open in various ways to release their seeds (**dehiscent fruits**) and those that remain closed (**indehiscent fruits**). Dry dehiscent fruits usually split open lengthwise, sometimes along one side only, as in the **follicle** of *Delphinium*, or along 2 sides, as in the **legume** of *Ulex* or *Cytisus*. More often, however, the fruit is a **capsule** which splits along the carpel margins (**septicidal dehiscence**) or along the midribs of the carpels (**loculicidal dehiscence**). In *Cerastium*, and some other genera of the Caryophyllaceae, the capsule splits at the apex along several lines, but only for a short way, so that a ring of triangular teeth is formed. In *Papaver* (Poppy), dehiscence is restricted to a ring of pores just below the sessile stigmatic disc, and in *Plantago* (Plantain) and *Anagallis* (Pimpernel) it takes the form of a transverse split round the capsule, the upper portion coming off like a lid. Plants in the Cruciferae have a particular kind of capsule, called a **siliqua** if long and a **silicula** if short, which opens from the base upwards. The 2 valves become detached, allowing the ripening seeds to fall from the central partition which remains joined to the receptacle. A few genera, such as *Ornithopus* (Leguminosae), produce a **lomentum**, a fruit formed from a single carpel which is constricted between the seeds and breaks up transversely into one-seeded portions. Dry indehiscent fruits are represented by the **cypsela** (Compositae), **caryopsis** (Gramineae),

and various kinds of **achenes**, **nuts**, and **nutlets** found in a number of different families. Some fruits are winged, e.g. the single **samara** in *Fraxinus* (Ash) and the double samara in *Acer* (Sycamore or Maple). Another form of dry fruit is the **schizocarp**, formed from 2 or more carpels, which breaks up into one-seeded portions (**mericarps**) when ripe, as in the Umbelliferae. Fleshy fruits occur in many families, and they are generally **berries** or **drupes**. Less common are the **pepo** (*Cucurbitaceae), **hip** (*Rosa*), and **pome** (*Malus* and related genera). In the genus *Rubus*, the fruit is a collection of small drupes (**drupelets**), and this is often termed an **aggregate fruit**. The flower of the strawberry, *Fragaria*, produces a **pseudocarp** (false fruit) consisting of the fleshy, much enlarged receptacle bearing achenes on its surface. The genera *Morus* (Mulberry), *Ficus* (Fig), and *Ananas* (Pineapple) have **multiple fruits** which are derived in various ways from the inflorescence. Because of their structure, which includes parts in addition to the fertilised ovary, they may also be termed 'false fruits'. Fruits are of primary importance in protecting the seeds as they mature and in aiding their dispersal. There is great variety in the forms that have evolved, and this variety has proved of considerable value to botanists concerned with classification.

FRUIT CLASSIFICATION

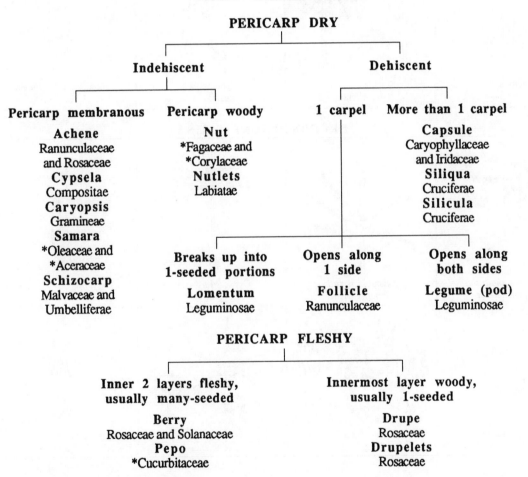

N.B. Pericarp = the fruit wall, derived from the ovary wall.
Fruits not formed by the ovary alone are called false fruits, some, e.g. Apple, Strawberry and Rosehip, derived from a single flower, and others, e.g. Fig, Mulberry and Pineapple, derived from an inflorescence.

DRY DEHISCENT FRUITS

Follicle *Consolida*
Cruciferae

Capsule *Papaver*
*Papaveraceae

Capsule *Cerastium*
Caryophyllaceae

Capsule *Anagallis*
Primulaceae

Siliqua *Arabis*
Cruciferae

Silicula *Capsella*
Cruciferae

Legume *Vicia*
Leguminosae

Lomentum *Ornithopus*
Leguminosae

DRY INDEHISCENT FRUITS

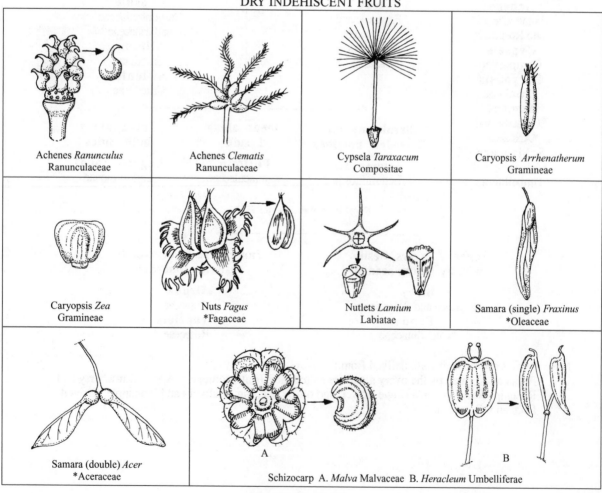

Achenes *Ranunculus*
Ranunculaceae

Achenes *Clematis*
Ranunculaceae

Cypsela *Taraxacum*
Compositae

Caryopsis *Arrhenatherum*
Gramineae

Caryopsis *Zea*
Gramineae

Nuts *Fagus*
*Fagaceae

Nutlets *Lamium*
Labiatae

Samara (single) *Fraxinus*
*Oleaceae

Samara (double) *Acer*
*Aceraceae

Schizocarp A. *Malva* Malvaceae B. *Heracleum* Umbelliferae

Fig. 25. Fruits (1).

FLESHY FRUITS

Drupe *Prunus* Rosaceae

Drupelets *Rubus* Rosaceae

Berry *Solanum* Solanaceae

Pepo *Cucurbita* *Cucurbitaceae

FALSE FRUITS

Pseudocarp *Fragaria* Rosaceae

Pome *Malus* Rosaceae

Hip *Rosa* Rosaceae

Multiple fruits A. *Ananas* *Bromeliaceae B. *Morus* *Moraceae C. *Ficus* *Moraceae

Fig. 25. Fruits (2).

Fig. 26. Leaf shapes (simple leaves)

Imparipinnate
(odd-pinnate)

Paripinnate
(even-pinnate)

Bipinnate

Palmate

Biternate

Trifoliolate

Fig. 27. Leaf shapes (compound leaves)

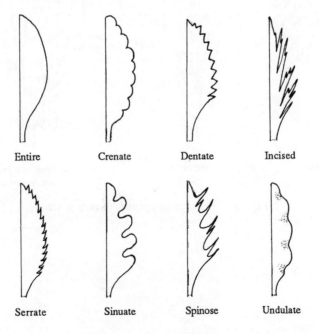

Entire

Crenate

Dentate

Incised

Serrate

Sinuate

Spinose

Undulate

Fig. 28. Leaf margins.

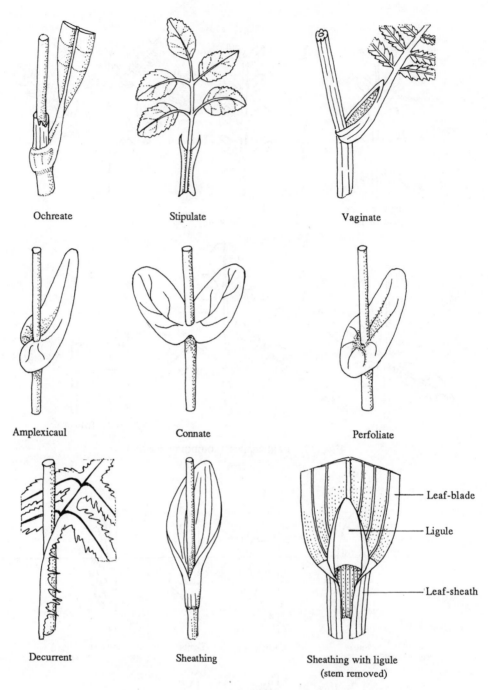

Fig. 29. Petioles and points of attachment.

Opposite	Alternate	Decussate	Whorled
Arenaria (Caryophyllaceae)	*Prunus* (Rosaceae)	*Syringa* (*Oleaceae)	*Lilium* (Liliaceae)

Fig. 30. Leaf arrangements.

Leaf spines of
Berberis (*Berberidaceae)

Stipular spines
(modified stipules) of
Robinia (Leguminosae)

Stem- and leaf-spines of
Ulex (Leguminosae)

Thorn

Simple leaf

Axillary bud

Compound leaf

Stipule

Prickle

Prickle

Thorns (modified stems) of *Crataegus*
(Rosaceae) with and without buds

Prickles (modified hairs) of *Rosa* (Rosaceae)

Fig. 31. Spines, thorns and prickles.

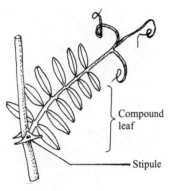

Compound leaf ending in
tendrils, *Vicia sativa* (Leguminosae)

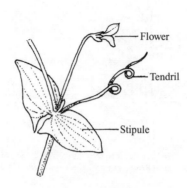

Leaf reduced to tendril and
stipules acting as leaves in
Lathyrus aphaca (Leguminosae)

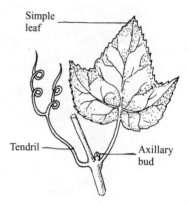

Branched tendril of Vine,
Vitis vinifera (*Vitaceae)

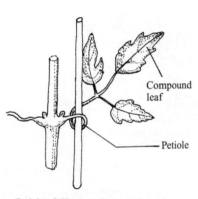

Petiole of *Clematis* (Ranunculaceae)
adapted for clinging to supports

Cladodes (modified stems)
of *Ruscus* (Liliaceae)

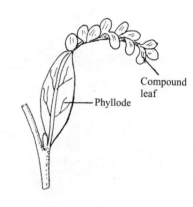

Phyllode (modified petiole)
of *Acacia* (Leguminosae)

Fig. 32. Tendrils, cladodes and phyllodes.

Simple (unicellular) hair
Capsella bursa-pastoris
(Cruciferae)

Glandular hair
Primula sinensis
(Primulaceae)

Multicellular hair
Olearia sp.
(Compositae)

Stellate hair
Alyssum corsicum
(Cruciferae)

Branched hair
Arabis hirsuta
(Cruciferae)

Fig. 33. Hairs.

The Plant Families

Nigella flower

Clematis flower

Nigella fruit (capsule)

Flowers, fruit and leaves of *Aquilegia vulgaris*, Columbine

Clematis fruit (achene)

Seed of *Helleborus foetidus* showing elaiosome

Fig. 34. Flowers and fruits of the Ranunculaceae.

Posterior perianth-segment

Spur of posterior perianth-segment

Spur formed by honey-leaves

Honey-leaf

Lateral perianth-segment

Anterior perianth-segment

Ovary surrounded by stamens

Anterior perianth-segment

Fig. 35. L.S. of flower of *Consolida ajacis*, Larkspur.

1 RANUNCULACEAE Juss. Buttercup family

58 genera and 1750 species

Distribution. World-wide.

General characteristics. Mostly herbaceous perennials, but some annuals, e.g. *Nigella*, and some woody climbers, e.g. *Clematis vitalba* (Traveller's Joy). Rootstock sometimes tuberous. Leaves usually alternate (opposite in *Clematis*) with sheathing bases and often very much divided; usually without stipules, but stipules present in *Trollius, Caltha, Thalictrum* and *Ranunculus*. Inflorescence variously formed; in *Anemone* and *Eranthis* a single terminal flower is produced, but more often a cymose branching occurs. In *Nigella* and *Anemone* there is sometimes an involucre of leaves below the flower. The flower-parts are free, variable in number, and spirally arranged upon a more or less elongated receptacle. The perianth may be distinguishable into calyx and corolla or it may be undifferentiated. The innermost perianth-segments often bear nectaries ('honey-leaves') or they may be reduced to small nectar-bearing sacs or scales. Flowers usually bisexual and regular (irregular in *Aconitum, Consolida* and *Delphinium*) with a superior ovary. Stamens usually numerous with the anthers facing outwards; staminodes sometimes present, e.g. *Aquilegia*. Carpels usually numerous and free, with either one basal or several marginal, anatropous ovules. (In *Nigella* the carpels are united, and in *Actaea* there is only one.) The stamens usually ripen before the stigma is receptive, and, as the anthers open, bend outwards from the centre. Fruit a group of achenes or follicles (capsule in *Nigella*, berry in *Actaea*). Seeds with small embryo and oily endosperm; elaiosome present in some species of *Helleborus*.

Economic and ornamental plants. The Ranunculaceae are mostly poisonous owing to the presence of alkaloids. Some species are or have been of medicinal value. Many have showy flowers and are grown in gardens. All the genera mentioned below are in cultivation.

Classification. A classification adapted from Hutchinson (see References) divides the family into 2 subfamilies and 5 tribes as follows:

I. Helleboroideae (carpels with more than 1 ovule; fruit a follicle or berry)
 1. Helleboreae (flowers regular)
 Aquilegia (70) N. temperate region.
 Trollius (31) N. temperate region.
 Helleborus (20) Europe, Mediterranean region, Asia.
 Nigella (14) Eurasia, Mediterranean region.
 Caltha (10) N. and S. temperate regions.
 Actaea (8) N. temperate region.
 2. Delphinieae (flowers irregular)
 Delphinium (200) N. temperate region.
 Aconitum (100) N. temperate region.
 Consolida (40) Mediterranean region to C. Asia.

II. Ranunculoideae (carpels with 1 ovule; fruit usually a group of achenes)
 3. Ranunculeae (leaves alternate, flowers without an involucre of leaves, calyx usually falling early, corolla usually present)
 Ranunculus (250) world-wide.
 Thalictrum (85) N. temperate region, tropical S. America, tropical and S. Africa.
 4. Anemoneae (leaves alternate, flowers with an involucre of leaves, calyx usually persistent and coloured, corolla absent)
 Anemone (120) almost world-wide.
 5. Clematideae (leaves opposite, calyx coloured, corolla absent)
 Clematis (230) N. and S. temperate regions, mountains of tropical Africa.

1.1 Ranunculus repens L. Creeping Buttercup

Distribution. Native in a wide range of habitats in the whole of the British Isles and also occurring as a common weed in gardens. It is found throughout most of Europe and N. Asia to Japan, and has been introduced into America and New Zealand.

Vegetative characteristics. A perennial herb with long, stout roots, and stolons which root at the nodes. The hairy leaves are divided into 3 lobes, each of which is again divided into 3 toothed segments.

Floral formula. K5 C5 A∞ G∞

Flower and inflorescence. The glossy, yellow flowers appear from May until August or September. The furrowed, hairy pedicels, 6 to 10 cm in length, support the flowers in a more or less erect position.

Pollination. The flowers of the Ranunculaceae are variously adapted to insect-pollination, e.g. *Clematis* (without nectar, visited by pollen-seeking short-tongued flies and bees); *Ranunculus* (nectar scarcely concealed, visited by hover-flies and small bees); *Nigella* (nectar in little closed cavities, visited chiefly by bees); *Aquilegia, Consolida* and *Delphinium* (nectar in long spurs, visited chiefly by bumble-bees). In *Ranunculus repens*, insects are attracted by the glossy, yellow petals and by the nectar that is secreted from the pocket-like nectaries at the base of the petals. The pollen from the mature stamens falls on to the petals, and is easily picked up by the visiting insects. After the pollen has been shed, the stigmatic surface of the carpels enlarges in order to receive pollen from another flower.

Alternative flowers for study. Ranunculus acris (Meadow Buttercup) which flowers from May to July, sometimes until October, and *R. bulbosus* (Bulbous Buttercup) which may begin to flower in March, but whose flowering period is normally May and June, are two of the more common species that are nearest to *R. repens* in floral structure.

A. A single flower detached from the cymose inflorescence. The corolla is composed of 5 free, glossy, golden yellow petals, and the calyx of 5 free, hairy sepals.

B. L.S. of the regular, bisexual flower with 2 of the 5 petals removed, exposing the numerous, free carpels surrounded by the numerous, spirally arranged stamens. The carpels and stamens are attached to a cone-shaped and hairy receptacle.
Sepals: 9 x 4 mm Petals: 16 x 12 mm

C. Detail of petal base, showing the pocket-like nectary situated on the upper surface.
Anthers: 2.25 mm Filaments: up to 5.5 mm

D. Details of stamens, which vary in length.
D1: Back view, showing the anther attached at its base to the filament.
D2: Front view of small stamen before dehiscence.
D3: Dehiscence of the 2-celled anther, showing the numerous pollen grains.

E. L.S. of a carpel containing a single ovule on a basal placenta. The stigma forms a beak-like structure which enlarges after the pollen has been shed from the stamens, favouring cross-pollination.

F. T.S. of a carpel, showing the basal placentation of the ovule.

G. The group of fruits (achenes) after the other floral parts have been removed. The achenes eventually dry and separate from the receptacle to aid fruit dispersal.
Achene (base to stigma): 2.35 mm
Ovary: 1 mm Ovule: 0.3 mm
Group of achenes: 4 x 4 mm

H. A carpel removed soon after fertilisation, showing the enlarged stigma. The carpel later develops into the dry, indehiscent fruit known as an achene.

1.1 *Ranunculus repens*, Creeping Buttercup

1.2 *Consolida ajacis*, Larkspur

1.2 Consolida ajacis (L.) Schur (Delphinium ajacis L.) Larkspur

Distribution. Native in the Mediterranean region, but locally naturalised elsewhere in Europe, including the British Isles. It seeds freely, and often escapes from gardens where it is grown as an ornamental plant. It was introduced into the U.S.A. and Canada, and is found naturalised from Nova Scotia to Minnesota southwards.

Vegetative characteristics. A softly hairy, annual herb, with an erect, simple or branched stem reaching a height of 1 m. The lower leaves are deeply palmately cut and have long petioles; the upper leaves are similar in shape but are more or less sessile.

Floral formula. P5 + (2) A∞ G1

Flower and inflorescence. The bright blue, irregular, bisexual flowers are borne in 4- to 16-flowered racemes in June and July. Occasionally the flowers may be purple, pink or white.

Pollination. Only long-tongued insects, mainly bumble-bees, can reach the nectar that is concealed in a spur formed by the 2 united honey-leaves. The stamens ripen first, so that an insect visiting the flower in its male stage becomes dusted with pollen. After shedding their pollen, the stamens wither and bend downwards, making room for the maturing style. A pollen-carrying insect that visits the flower in its female stage transfers the pollen to the receptive stigma. In this way cross-pollination is accomplished.

Alternative flowers for study. Other suitable species of *Consolida* that are often cultivated include *C. orientalis* and *C. regalis*. These and other annual species with a single carpel are placed in the genus *Consolida* to distinguish them from species of *Delphinium* which have 3-5 carpels.

A. The irregular, bisexual flower in its natural position. 4 of the 5 perianth-segments (sometimes referred to as petaloid sepals) can be seen. The upper (posterior) perianth-segment is prolonged at its base into a spur. The 2 united honey-leaves, visible at the centre of the flower, are similarly prolonged, the spur thus formed fitting into the perianth-spur and protected by it.

B. L.S. of the flower. The reproductive organs, which are shielded on their upper side by the erect, lobed limb of the 2 united honey-leaves, are situated near the entrance to the tapering, nectar-secreting spur. Parts of the upper perianth-segment, which forms the enclosing spur, and 3 of the 4 unspurred segments are shown.
Upper perianth-segment (limb): 18 x 10 mm
(spur): 16-17 mm
Unspurred perianth-segments:
17-18 x 10-20 mm
Limb of honey-leaves: 17-18 x 5-6 mm

C. Some of the numerous stamens that are arranged spirally in 5 series round the superior ovary. The reproductive organs are shown here in an erect position though they are horizontal in nature. The anthers, some of which have reached maturity, are 2-celled and dehisce lengthwise. The scars on the receptacle show where perianth-segments have been removed.
Stamens: *c*. 9 mm

D. Detail of a single stamen, showing the filament broadening considerably towards the base. The anther is attached at its base to the filament.
Filaments: 6-8 mm Anthers: 1.75 x 1.25 mm

E. L.S. of the superior ovary ending in a short, curved style. The ovary is softly hairy and contains numerous ovules in the single loculus. Round the base of the ovary are portions of the perianth-segments.
Ovary (at flowering-time): 6 mm

F. T.S. of the ovary, showing 3 of the numerous ovules attached to the marginal placenta.
Ovary: 2 mm in diameter.

G. The fruit prior to dehiscence. When ripe, the fruit (a follicle) becomes erect and splits down the seam (suture) that is on its ventral side, liberating the numerous black seeds.
Fruit: 20 x 5 mm

1.3 **Helleborus foetidus** L. Stinking Hellebore

Distribution. Native in S. and W. Europe eastwards to Syria. In the British Isles it occurs locally in woods and scrub on shallow, calcareous soils. It is sometimes cultivated in gardens, occasionally escaping and becoming naturalised.

Vegetative characteristics. A perennial herb with a stout, overwintering, branched stem reaching a height of between 20 and 80 cm. The stem bears large leaves, deeply divided into 3-9 toothed segments.

Floral formula. K5 C5-10 A ∞ G(2-5)

Flower and inflorescence. Several drooping, green flowers, often with reddish-purple edges to the perianth-segments, are borne in a branched cyme between January and the middle of April.

A. Portion of the inflorescence, backed by an illustration of a fully developed, lower stem-leaf. The bell-shaped or globular flowers have pedicels with prominent pale green bracts. Note the leaf-sheath on the main stem which ends in a vestigial leaf-blade.

B. A young flower. The free sepals forming the calyx are in 2 whorls, the outer consisting of 2, and the inner of 3 sepals. Within the calyx are the reproductive organs; some of the stamens are just protruding from the calyx.
Outer sepals: up to 18 x 13 mm
Inner sepals: up to 14 x 13 mm

C. A flower with sepals removed to reveal the cluster of stamens prior to dehiscence. Surrounding the stamens and attached to the receptacle are 4 of the 5-10 green nectaries (sometimes called honey-leaves) which are considered to be modified petals.
Cluster of stamens + nectaries: 12 x 8 mm

D. A single nectary. The nectary is tubular in shape, broadening out at the top and tapering at the base where it joins the receptacle. The sculpturing of the rim is very variable.
Nectary: 7 x 2 mm

E. A flower with part of the calyx cut away and most of the stamens and nectaries removed to reveal the prominent receptacle which bears 30-55 stamens. At the apex of the receptacle is the superior ovary composed of 3 carpels that are united at the base (one of these has been removed for clarity). The number of carpels can vary

Pollination. The abundance of nectar and pollen, and the foetid smell attract the few long-tongued insects that are active during the flowering season (mainly bees and hoverflies). The stigmas ripen first and curve outwards in the mouth of the calyx. Later on, the stamens lengthen and shed their pollen, the outer ones ripening first, followed by those nearer the centre of the flower.

Alternative flowers for study. Any other species or cultivar of *Helleborus* would be suitable, but in the interests of conservation only garden grown plants should be chosen. *Eranthis hyemalis* (Winter Aconite) has usually 6 yellow sepals, 6 nectaries, and 6 stalked carpels.

between 2 and 5. At the left of the flower is one of the tube-shaped nectaries.

F. Detail of upper portion of stamen. F1: Ventral view of anther. One of the pollen sacs is dehiscing lengthwise to shed its pollen. F2: Dorsal view of anther, showing the extension of the filament (connective) that joins the 2 pollen sacs together.
Anthers: 2-2.5 x 1 mm
Filaments: up to 12 mm

G. A single carpel. The ovary bears a long style ending in a capitate stigma. On the left of the base of the ovary can be seen the point of attachment to the other carpels.
Ovary: 4.5-5 x 1.5-2 mm Style: 6-7 mm

H. L.S. of ovary. The ovules are arranged in 2 rows along the placenta, near to the ventral seam (suture).

I. T.S. of ovary with marginal placentation. Note that the 2 ovules attached to the placenta are in the same plane. The groove forming the ventral suture can be clearly seen. Glandular hairs are present on the young ovary.
T.S. of ovary: 1.5 mm

J. The fruit. At maturity the carpels separate from one another to form individual follicles which dehisce down the ventral suture to disperse the seeds. The persistent calyx remains at the base of the follicles.
Follicle: *c.* 25 x 7 mm

2.1 *Cerastium tomentosum*, Snow-in-summer

2 CARYOPHYLLACEAE Juss. Pink family

89 genera and 2070 species

Distribution. World-wide, especially northern temperate and warm regions.

General characteristics. Mostly herbs, the leaves opposite, simple, usually entire and without stipules, the stem often forked and swollen at the nodes. The inflorescence usually terminates the main axis and is typically a form of cyme. Flowers usually bisexual and regular. Epicalyx present in some genera (e.g. *Dianthus*). Calyx of 4 or 5 overlapping sepals, free or united at base. Corolla usually of 4 or 5 free petals, often differentiated into limb and claw. The apex of the petals is often notched or deeply divided. Stamens usually 5 or 10, ripening before the stigmas. Ovary superior, of 2-5 united carpels, unilocular at least above, with free-central or basal placentation. Styles free, as many as carpels. Ovules usually many, in double rows corresponding to the carpels, sometimes few or 1, usually campylotropous. The ovary, stamens and corolla are sometimes borne on an elongated column (e.g. *Silene, Lychnis*). At the base of the ovary traces of the septa may be visible. Fruit usually a capsule containing several or many seeds, sometimes an indehiscent, 1-seeded nutlet. In most cases the capsule opens by splitting from the apex into teeth which bend outwards, leaving an opening. The splitting may take place in as many, or in twice as many, lines as there are carpels. The seeds cannot escape from the capsule until shaken by wind or animals when, being small and light, they have a good chance of being widely distributed. Rarely (e.g. *Cucubalus*) the fruit is berry-like, but dry and indehiscent when ripe. Embryo usually curved round the perisperm, but in a few cases nearly straight.

Economic and ornamental plants. The family contains some of our most popular and fragrant decorative plants, e.g. *Dianthus barbatus* (Sweet William), *D. caryophyllus* (Carnation), and other species of *Dianthus* from which the Garden Pinks have been derived; various species of *Silene, Saponaria*, and *Lychnis*, including *L. chalcedonica* (Maltese Cross), *Gypsophila paniculata* and its cultivars, *Cerastium tomentosum*, and species of *Arenaria* and *Petrocoptis*.

Classification. The following classification, based on that in *Flora Europaea*, Tutin *et al.* (see References), divides the family into 3 subfamilies, all of which secrete nectar at the base of the stamens. However, only long-tongued insects, mainly bees and Lepidoptera, can reach the nectar of the Silenoideae since the calyx-tube excludes short-tongued insects. In the other subfamilies, the sepals are not united into a tube, and the flowers open wide, so that the nectar is available to all kinds of insects.

I. Alsinoideae (stipules absent; sepals free or joined only at base)
 Arenaria (150) N. temperate region.
 Stellaria (120) world-wide.
 Cerastium (60) almost world-wide.
 Sagina (25) N. temperate region, tropical mountains.

II. Paronychioideae (stipules present; sepals free)
 Paronychia (50) world-wide.
 Drymaria (50) mainly Central and S. America and W. Indies.
 Spergularia (40) world-wide.
 Herniaria (20) Europe, Mediterranean region to Afghanistan, S. Africa.
 Corrigiola (10) world-wide.
 Spergula (5) temperate regions.

III. Silenoideae (stipules absent; sepals joined in a tubular or bell-shaped calyx)
 Silene (500) N. temperate region.
 Dianthus (300) Europe, Asia and Africa.
 Gypsophila (125) temperate Eurasia, Egypt.
 Lychnis (35) N. temperate and arctic regions.
 Saponaria (30) temperate Eurasia.

2.1 Cerastium tomentosum L. Snow-in-summer

Distribution. A commonly grown rock-garden and wall plant native in the Caucasus and S.E. Europe. It often escapes from cultivation, and has now become established in a number of places in the British Isles and N. America.

Vegetative characteristics. A perennial, mat-forming herb with stems and leaves densely covered with white hairs. The leaves are linear-lanceolate with slightly revolute margins.

Floral formula. K5 C5 A10 G(5)

Flower and inflorescence. The showy, white flowers, forming a loose cyme, are borne at the end of the stems from May until September.

Pollination. The flowers of *C. tomentosum* open wide, enabling even short-tongued insects to reach the nectar that is secreted by glands at the base of the stamens. The stamens normally ripen before the stigmas are receptive, favouring cross-pollination.

Alternative flowers for study. The Common Mouse-ear Chickweed, *C. fontanum* ssp. *triviale* (*C. holosteoides*), which has a similar flowering period, is abundant throughout the British Isles. It differs little in floral structure from *C. tomentosum* but is perhaps less suitable because of its smaller flowers.

A. The regular, bisexual flower with 2 of the 5 free, notched petals removed to expose the reproductive parts. The 5 free, hairy sepals allow the petals to open wide, in contrast to those flowers in the family where the sepals are united into a calyx-tube. The superior ovary is surrounded by the free stamens which are in 2 whorls of 5. Nectar is secreted at the base of the stamens.
Corolla: 12-18 mm in diameter
Sepals: *c.* 7 mm Petals: *c.* 27 mm

B. A long and a short stamen attached to a teased-out portion of the receptacle. The longer one belongs to the outer whorl of stamens which stand opposite the petals; the shorter one belongs to the inner whorl of stamens which stand opposite the sepals. The anthers are 2-celled, face inwards, and dehisce lengthwise.
Anthers: 1.5 mm Filaments: 6-7 mm

C. Sepals, petals and stamens have been cut back to reveal the superior ovary consisting of 5 united carpels and bearing 5 free styles.
Ovary: *c.* 2 mm Styles: 4-5 mm

D. Detail of upper portion of style, showing the papillose stigma.

E. A single papilla cut from the stigmatic surface.

F. T.S. of the ovary, showing the numerous ovules in the single loculus with free-central placentation.
Ovary (at flowering time): 1.75 mm in diameter

G. L.S. of ovary. The ovules are arranged in double rows which correspond to the number of united carpels.

H. The fruit is a capsule which dehisces at the apex by 10 small teeth (twice the number of carpels). When the capsule is shaken by the wind or by animals, the numerous small seeds are dispersed. In the genus *Cerastium*, the capsule is always longer than the sepals.
Capsule: *c.* 10 x 4 mm Teeth: 1.25 mm

2.2 **Silene dioica** (L.) Clairv. Red Campion

Distribution. Native throughout most of Europe, also in N. Africa, and western and central Asia. It is locally abundant in the British Isles, and is found in woods and hedgerows on moist but well-drained soils. Dwarf forms occur on exposed rocks and on mountains.

Vegetative characteristics. A perennial herb with a slender, creeping rootstock that produces both decumbent non-flowering shoots and erect, flowering stems to a height of 30-90 cm. The stems are covered with soft, spreading hairs and are sometimes slightly sticky. The hairy leaves are arranged in pairs and their stalks are of varying length.

Floral formula. Male: K(5) C5 A10 G(rudimentary)
 Female: K(5) C5 A(rudimentary) G(5)

Flower and inflorescence. The rose-coloured, red, or rarely white, unisexual flowers are borne on separate plants during May and June. The inflorescence is in the form of a many-flowered dichasial cyme.

Pollination. Long-tongued bumble-bees, hoverflies and butterflies visit the flowers to seek out the nectar secreted by the prominent nectary situated at the base of the ovary or stamens. Since male and female flowers occur on different plants cross-pollination is ensured.

Alternative flowers for study. In the absence of *S. dioica*, *S. alba* (White Campion) and *S. vulgaris* (Bladder Campion) would be suitable, both species usually having unisexual flowers. The genus *Silene*, which has 3 or 5 styles, may be confused with *Lychnis*, which always has 5 styles. In *Lychnis*, however, the capsule dehisces by 5 teeth, while in *Silene* the capsule opens by twice as many teeth as styles.

A. Part of the inflorescence of a male flower. The calyces, pedicels, leaves and stem are covered with soft, spreading hairs. The apex of the petal is deeply notched. In the male flower, the calyx-tube is cylindrical with 10 veins, while in the female flower the calyx-tube is ovoid with 20 veins.
 Flower: 26 mm in diameter
 Calyx (male): 13-14 x 5 mm
 (female - not shown): 14 x 7 mm

B. One of the petals removed from the receptacle. Note the prominent coronal scale at the base of the coloured limb before it narrows into the more or less colourless claw.
 Limb of petal: 12 x up to 10 mm
 Claw of petal: 15 mm (basal part not shown)

C1. Male flower at an early stage of flowering. Part of the perianth has been cut away to show the 10 stamens of different lengths. These surround a rudimentary carpel that is embedded in the nectary but attached to the receptacle.

C2. Detail of the base of a male flower showing the rudimentary carpel at the centre of the ring-like nectary.
 Nectary: 1.5 mm in diameter
 Rudimentary carpel: 1.5 mm

D. Detail of anther attached by its back to the filament. The anthers face inwards and dehisce lengthwise.
 Anther: 1.5 mm
 Filament: up to 18 mm

E. L.S. of female flower. The ovoid, hairy calyx surrounds the lower part of the free petals. Three of the 5 styles can be seen. The superior ovary contains numerous ovules attached to the prominent free-central placenta.

F. T.S. of ovary. The ovary is formed from 5 united carpels, as is indicated by the remains of 5 rudimentary partitions (septa) on the inner surface of the ovary.
 Ovary: 3 mm in diameter

G. Female flower, with perianth removed to reveal the superior ovary bearing 5 styles that project well beyond the mouth of the corolla.
 Ovary: 7 x 2.5 mm
 Styles: up to 10 mm

H. Lower portion of the ovary. The calyx has been removed to show the point of insertion of the claws of the 5 free petals. Between 2 of the claws can be seen one of the rudimentary stamens apparently attached to the ring-like nectary.

I. The fruit, a globose to broadly ovoid capsule that dehisces by 10 recurved teeth. Part of the calyx has been removed for clarity.
 Capsule: 12 x 10 mm
 Opening at top of capsule: 6 mm in diameter

A

B

I

C1

D

E

G

C2

F

H

3.1 *Malva sylvestris*, Common Mallow

3 MALVACEAE Juss. Mallow family

121 genera and 1550 species

Distribution. World-wide, especially tropical regions.

General characteristics. Herbs, shrubs or trees with alternate, often palmately lobed leaves. Stipules usually present. Flowers solitary or in cymes, bisexual, regular, with the flower-parts usually in fives. Epicalyx often present. Calyx of 5 free or united sepals, valvate. Corolla of 5 free, usually asymmetrical petals, twisted or overlapping in bud. Stamens usually many owing to branching of the inner whorl (outer whorl usually absent), all united below into a tube which is joined at the base to the petals; anthers one-celled, pollen grains spiny. Ovary superior, of 1 to many united carpels (frequently 5), multilocular, with axile placentas; ovules 1 to many in each carpel, anatropous, usually ascending, sometimes pendent. *Malvaviscus* has a berry, but the rest of the family have dry fruits, either capsules or schizocarps. Embryo usually curved; endosperm oily, copious to scanty or absent.

Economic and ornamental plants. The most important genus commercially is *Gossypium* (Cotton), several species of which have contributed to the modern hybrid cultivars. Cotton is manufactured from the dense mass of long hairs attached to the seeds. *Hibiscus esculentus* (Okra) is widely grown as a foodstuff in the tropics for its fruit or 'pods', which are picked while they are still immature. Other members of the genus *Hibiscus* are cultivated for their showy flowers, while species of the hardier genera *Malva* (Mallow), *Lavatera*, *Sidalcea*, *Althaea*, and *Alcea* (Hollyhock) are often found in gardens in temperate regions.

Classification. The following classification into 4 tribes is based on that by Schumann (see References):

A. Carpels in several whorls, forming vertical rows.
 1. Malopeae
 Malope (3) Mediterranean region.

B. Carpels in one whorl.
 2. Malveae (schizocarp; styles as many as carpels)
 Abutilon (100) tropics and warm regions.
 Alcea (60) Mediterranean region to central Asia.
 Malva (30) Eurasia, Mediterranean region.
 Lavatera (25) Canary Islands, Mediterranean region to C. Asia, Australia, California.
 Althaea (12) Europe to N.E. Siberia.
 Anoda (10) tropical America.
 3. Ureneae (schizocarp; styles twice as many as carpels)
 Pavonia (150) tropics and warm regions.
 Urena (6) tropics and warm regions.
 4. Hibisceae (capsule)
 Hibiscus (200) tropics and subtropics.
 Gossypium (39) tropics and subtropics.

Fig. 37. Fruit of *Hibiscus esculentus*, Okra, whole and in section.

Capsule with calyx Capsule at dehiscence with cotton boll Seed with hairs (cotton) attached

Fig. 36. Fruit and seed of *Gossypium*, Cotton.

49

3.1 Malva sylvestris L. Common Mallow

Distribution. Native throughout most of Europe including the British Isles, and commonly found growing on waste places and roadsides. It has become naturalised in the central and eastern parts of N. America.

Vegetative characteristics. A perennial herb with hairy, long-stalked leaves, heart-shaped at the base, and with 5-7 shallow lobes.

Floral formula. K(5) C5 A(∞) \underline{G}(∞)

Flower and inflorescence. The flowers arise in clusters in the axils of the leaves. They are large, bisexual, and regular, with a tendency to asymmetry, and make an attractive show from June to September with their rose-purple, notched petals. Below the calyx of each flower is an epicalyx consisting of 3 free segments.

Pollination. Many kinds of insects are attracted by the showy petals and by the nectar which is secreted at their base. In the first stage of flowering, the ripening anthers form a pyramid at the centre of the flower above the immature stigmas. After the stamens have shed their pollen they bend back, exposing the developing stigmas which spread out and, in their turn, become the dominant feature at the centre of the flower.

Alternative flowers for study. M. sylvestris is a readily available species, but any other member of the genus would be suitable. Glasshouse grown species of *Hibiscus*, which have similar flowers, may be available at other times of the year.

A. A flower-bud, showing the calyx and, below it, the 3-lobed epicalyx, that is variable in size and shape.
Calyx + epicalyx: 6 mm

B. The regular, bisexual flower, with one of the 5 petals removed to reveal the withering anthers and, above them, the protruding stigmas. The 5 petals are twisted in bud, and at their base are fine hairs which protect the nectar-secreting pits. The 5 hairy sepals forming the calyx are united in their lower half. Below the calyx can be seen part of the epicalyx, the 3 lobes of which are about one-third shorter than the calyx.
Corolla: *c.* 32 mm in diameter Petal: 17 mm
Calyx-lobe: 3 x 3 mm

C. L.S. of centre of flower. The tubular group of stamens surrounds the column of styles which arises from the centre of the flower. At the top, the developing stigmas are taking the place of the withering anthers.
Staminal tube: 8-9 mm Ovary: 1.25 mm

D1. The free, upper portion of 2 of the numerous stamens whose filaments are united below into a tube. The one-celled, kidney-shaped anthers dehisce by a slit across the top, turning outwards to shed their characteristically large, round, spiny pollen grains.
Free portion of stamens: 1-2 mm
Anthers: 0.5 x 0.5-0.75 mm

D2 & 3. Anther before and at dehiscence. The anther is attached by its back to the filament.

E. A portion of the central column formed by the united styles ending in the free stigmas, which are equal in number to the carpels.

F. T.S. of the superior ovary, showing the numerous loculi. Each loculus contains a single ovule with axile placentation.
Ovary (at flowering time): 2 mm in diameter

G. The fruit, a schizocarp, which separates into many indehiscent one-seeded nutlets surrounded by the persistent calyx and epicalyx.
Fruit: 7 mm in diameter

H. A single, wedge-shaped nutlet showing the reticulate markings on its back. When the nutlets are ripe, they break away from each other and from the persistent central column formed by the apex of the receptacle.
Nutlet: 3 x 3 mm

4.1 *Salix caprea*, Goat Willow

4 SALICACEAE Mirbel Willow family

2 genera and 335 species

Distribution. Chiefly N. temperate region.

General characteristics. Trees, shrubs and sub-shrubs with male and female flowers on separate plants. Leaves simple, usually alternate. Stipules usually present, but often soon falling. Flowers in catkins, each flower with a scale or bract at its base. Perianth absent or represented by a cup-shaped disc or small nectary. Stamens 2 to many, free or united, anthers 2-locular. Ovary superior, of 2-4 united carpels, unilocular, more or less flask-shaped, with numerous anatropous ovules on 2-4 parietal or basal placentas. Fruit a 2- to 4-valved capsule; seeds with silky hairs arising from the funicle; embryo straight, in little or no oily endosperm.

Economic and ornamental plants. Both genera in this family are of economic importance for their wood which is used for a variety of purposes. The bark of some species of *Populus* (Poplar) and *Salix* (Willow) is of medicinal value, and the slender twigs of some Willows are in demand for basket-making. Many hybrids and cultivars of both genera are in existence. Weeping forms of *Salix* are especially popular as decorative trees in damp situations, and *Populus nigra* 'Italica' (Lombardy Poplar) has been a familiar avenue tree for over 200 years.

Classification.

A. Bud covered by a single scale, nectaries usually 1-4, stamens usually 2, 3, or 5, bract entire, leaves usually short-stalked with narrow blades. *Salix* (300) chiefly N. temperate region.

B. Bud covered by several overlapping scales, nectary absent, stamens 4 to many, bract toothed or laciniate, leaves usually long-stalked with broad blades.
Populus (35) N. temperate region.

Bract of male flower that overlaps the disc

Female catkin

Male flower: stamens arising from a disc

Male catkin

L.S. of female flower

Female flower: carpel with a prominent stigma

Fig. 38. Flower structure of *Populus* (Poplars are all wind-pollinated).

4.1 Salix caprea L. Goat Willow

Distribution. Common in woods, hedges, scrubland, and the margins of ponds, lakes, and rivers throughout the British Isles and most of Europe. Native also in temperate Asia eastwards to Turkestan. It is much cultivated in N. America, and has become naturalised in parts of the N.E. United States.

Vegetative characteristics. A deciduous shrub or small tree with alternate, ovate to obovate leaves, dark green above and with grey hairs on the under surface.

Floral formula. Male: P0 A2
Female: P0 G̱(2)

Flower and inflorescence. The unisexual flowers, without either sepals or petals, are solitary in the axil of a scale or bract. They are arranged spirally round a central axis, and collectively form the type of inflorescence known as a catkin. At the base of the catkins, which reach maturity in March and April, are small bracts.

Pollination. The genus *Salix* is pollinated by insects in contrast to the related genus *Populus* which is wind-pollinated. Various kinds of insects, especially early bees, are attracted to the catkins at a time of the year when there are relatively few sources of nectar available. All the catkins are made conspicuous by appearing before the leaves, and the male catkins particularly so by the presence of bright yellow pollen at the tips of the stamens. Both male and female flowers produce abundant nectar from the well-developed nectar-glands at their base. Since male and female flowers are on separate trees, cross-pollination within the species is assured, but there is also frequent hybridisation between different species.

Alternative flowers for study. Any species of *Salix* would be suitable since the structure of the flowers is fundamentally similar. The main differences between the species occur in time of flowering, dimensions of the catkins, colour of the scales, and also in the number of stamens, which may vary from 2 to 5 or more.

A. A single female flower, composed of 2 united carpels, on a short stalk. Joined to the stalk is an entire, hairy bract, at the base of which is a prominent nectary.
Bract: 2-3 mm Ovary: 4 mm

B. L.S. of the superior, unilocular ovary, which contains numerous ovules.

C. T.S. of ovary, showing the 2 parietal placentas with ovules attached.
T.S. of ovary (at flowering time):
1.25 x 0.75 mm

D. Detail of style, branched at the apex to form 2 conspicuous, stigmatic surfaces.
Branch of style: 0.5 mm

E. A single male flower, composed of 2 free stamens. The entire, hairy bract has a prominent nectary at its base. (In some species the stamens are joined at their base.)
Bract: 2 mm
Stamens (in young flower): 4-5 mm
(in mature flower): 10 mm

F. A stamen prior to dehiscence. The anther dehisces lengthwise and sheds abundant pollen.
Anther: 1.25 mm

G. An immature male catkin. Its hairy appearance is due to the closely packed, hairy bracts at the base of the flowers.
Male catkins (at maturity): 19-20 mm
Female catkins (illustrated on p. 52): 25 mm

H. Detail of the solitary nectary situated at the base of both male and female flowers.
Nectary: 1 mm

I. The fruit at dehiscence, a capsule which opens by 2 valves to liberate the numerous seeds.
Fruit: 6 mm

J. The seed, bearing long hairs that arise from the funicle and aid wind-dispersal. The seeds quickly become dehydrated owing to the thin seed-coat and early germination is essential.
Seed: 1.5 mm Hairs: 7 mm

Lunaria annua, Honesty.
Flat, thin-walled silicula
at dehiscence.

Capsella bursa-pastoris, Shepherd's Purse.
Siliculae before and after dehiscence.

Draba incana, Hoary Whitlow-grass. Twisted
siliqua at early dehiscence, and septum
showing seed arrangement.

Raphanus raphanistrum, Radish.
Lomentum-like fruit, with portion cut open
to reveal the more or less spherical seeds.

Erysimum cheiri, Wallflower.
Flattened siliqua at dehiscence.

Sinapis alba, White Mustard.
Siliqua with long beak.

Isatis tinctoria, Woad.
Pendulous, indehiscent,
1-seeded, winged fruit.

Coronopus didymus, Lesser Swine-cress.
Fruit with reticulate-pitted surface
which splits into 2 1-seeded portions.

Thlaspi arvense, Field Penny-cress.
Silicula with 2 broadly winged
valves and a distinct apical notch.

Bunias orientalis, Warty Cabbage.
Warty, 1- or 2-seeded
indehiscent fruit.

Fig. 39. Fruits of the Cruciferae.

5 CRUCIFERAE Juss. (Brassicaceae Burnett)

390 genera and 3000 species Mustard family

Distribution. World-wide, chiefly N. temperate region, especially Mediterranean.

General characteristics. The family consists of annual to perennial herbs or rarely small shrubs. Leaves simple, usually alternate, with unicellular simple or branched hairs. Stipules absent. Inflorescence usually a raceme or corymb and nearly always without bracts or bracteoles. Flowers usually regular, bisexual. Calyx of 4 free sepals in 2 whorls. Corolla of 4 free petals, alternate with the sepals, usually spreading in the form of a cross and often elongated into a claw at the base. Stamens in 2 whorls, the outer whorl consisting of 2 short stamens, the inner of 4 long ones (tetradynamous); anthers facing inwards. Ovary superior, composed of 2 united carpels with parietal placentas, but 2-locular on account of a central partition (replum) formed by the union of outgrowths from the placentas; stigma capitate to bilobed, on a short style; ovules usually campylotropous. Fruit a pod-like capsule, called a siliqua if it is at least 3 times as long as broad, e.g. *Arabis caucasica* (see below) and *Erysimum cheiri* (Wallflower), and a silicula if it is shorter, e.g. *Capsella bursa-pastoris* (Shepherd's Purse) and *Lunaria annua* (Honesty). When the fruit dehisces, the valves break away from below upwards, leaving the partition with the seeds still attached to it. The fruit may be flattened in 2 ways, either parallel or at right angles to the replum; this character is of systematic importance. Sometimes the fruit may be jointed between the seeds as in a lomentum, e.g. *Raphanus sativus* (Wild Radish). Achene-like, one-seeded fruits occur in a few genera, e.g. *Isatis tinctoria* (Woad). Seed characters are also of systematic importance. The seed has a large, oily embryo but little or no endosperm and the seed-coat is often mucilaginous.

Economic and ornamental plants. Many of the Cruciferae are cultivated for their food value. These include varieties of *Brassica oleracea* (Cabbage etc.), *B. rapa* (Turnip), *Sinapis alba* (Mustard), *Rorippa nasturtium-aquaticum* (Watercress), *Lepidium sativum* (Garden Cress), and *Armoracia rusticana* (Horse-radish). *Isatis tinctoria* (Woad) was formerly cultivated as the source of a blue dye. Many others are commonly grown for decorative purposes. In addition to those already mentioned, some of the best known are *Matthiola* and *Malcolmia* (Stocks), *Iberis* (Candytuft), *Lobularia maritima* (Sweet Alison), *Aubrieta*, and *Alyssum*.

Classification. Grouping within the Cruciferae has proved difficult, and many classifications have been devised. The following list shows some of the larger genera:

Draba (300) N. temperate and arctic regions, mountains of Central and S. America.
Alyssum (168) Mediterranean region to Siberia.
Lepidium (150) world-wide.
Cardamine (130) world-wide, mainly N. hemisphere.
Arabis (120) N. temperate region, Mediterranean to tropical African mountains.
Erysimum (80) Mediterranean region, Europe, Asia.
Sisymbrium (80) Eurasia, Mediterranean region, S. Africa, N. America, Andes.
Heliophila (71) S. Africa.
Rorippa (70) almost world-wide.

Fig. 40. *Draba aizoides*, a yellow-flowered alpine.

Fig. 41. The irregular flower of *Iberis*, Candytuft, an uncommon feature in Cruciferae.

5.1 Arabis caucasica Schlecht. Garden Arabis

Distribution. A native of mountains of the Mediterranean region and Near East to Iran. It was introduced into the British Isles as an ornamental plant, and is often grown in rock gardens and on walls, sometimes escaping from cultivation and becoming naturalised.

Vegetative characteristics. A perennial mat-forming herb having greyish green or whitish leaves with 2-5 blunt teeth along each margin. The stem-leaves have small, rounded or pointed lobes at their base, while the basal leaves narrow gradually into a short petiole.

Floral formula. K2+2 C4 A4+2 G(2)

Flower and inflorescence. The clusters of white flowers form more or less erect racemes. The main flowering period is from March until June, but there may be a second show of flowers later in the year.

Pollination. As in most members of the Cruciferae, nectar secreted into the bases of the pouched sepals attracts insects and promotes cross-pollination. However, self-pollination of the flowers is also a frequent occurrence.

Alternative flowers for study. The smaller-flowered species *Arabis alpina* (Alpine Rock-cress), *Sinapis arvensis* (Charlock), and *Erysimum cheiri* (Wallflower) are suitable alternatives which flower in the late spring and early summer. Interesting variations in the flower are exhibited by some genera, e.g. in *Iberis*, the flowers are irregular, the 2 outer petals being much larger than the 2 inner ones (see Umbelliferae); in *Capsella bursa-pastoris* (Shepherd's Purse), the petals are occasionally lacking, and in *Coronopus didymus* (Lesser Swine-cress), *Lepidium ruderale* (Narrow-leaved Pepperwort), and *L. densiflorum* the petals are often entirely absent.

A. A flowering shoot with rooting stem detached from the parent plant.

B. View of the flower from above, showing the 4 free petals which alternate with the 4 free sepals.

C. L.S. of the flower, showing 2 of the 4 petals and sepals. The free petals are clawed and the inner sepals are pouched (saccate) at the base. Surrounding the ovary are the tetradynamous stamens, typical of the family. Both of the short stamens and 2 of the long ones are illustrated.
Sepals: 5 mm Petals: 12mm Petal-claw: 7 mm
Long stamens: 8 mm Short stamens: 6 mm

D. Petals and sepals have been removed to expose 2 of the long stamens and the superior ovary. The anthers are attached at the back to the filaments and face inwards. They consist of 2 cells that dehisce lengthwise. At the base of the filaments and joined to the receptacle are nectar-secreting glands. The very short style has a knob-like stigma which matures at the same time as the stamens.
Ovary + stigma (at flowering time): 9 mm
Stigma: 0.35 x 0.85 mm

E. L.S. of the ovary, showing the 2 carpels with numerous ovules.

F1 & 2. T.S. of the ovary, cut in 2 different ways to show the parietal placentation of the ovules.
Ovary: 0.5 mm in diameter

G. The fruit, a siliqua, formed from the fertilised carpels. When the siliqua dehisces, the 2 valves open upwards from the base, exposing the slightly winged seeds arranged along each of the placentas.
Fruit: 25 x 2 mm

A

B

C

D

F1

F2

E

G

6.1 *Erica herbacea*, Winter Heath

6 ERICACEAE Juss. Heath family

103 genera and 3350 species

Distribution. World-wide.

General characteristics. Mostly evergreen shrubs, rarely trees (e.g. *Arbutus*). Leaves simple, often leathery, usually alternate, sometimes opposite or whorled, without stipules. Inflorescence often surrounded at the base by numerous bracts. Pedicels usually bearing 2 or 3 bracteoles. Flowers usually bisexual and regular (irregular in *Rhododendron*). Calyx of usually 4 or 5 united sepals, persistent. Corolla of usually 4 or 5 united petals, urn-shaped, bell-shaped, or salver-shaped. Stamens usually 8 or 10, free; anthers 2-locular, becoming inverted during development so as to appear to face inwards when mature, usually dehiscing by terminal pores, often with appendages; pollen in tetrads. Nectar-secreting disc usually present. Ovary usually superior, of usually 4 or 5 united carpels, the loculi opposite the corolla-lobes, with 1 to many anatropous ovules in each loculus; placentation usually axile. Style simple, stigma usually capitate. Fruit a capsule, drupe or berry; seeds usually numerous; embryo embedded in copious oily endosperm.

Economic and ornamental plants. The fruits of a number of species of *Vaccinium* are edible, being variously known as Cranberries, Bilberries, Blueberries, and Cowberries. Species of the American genus *Gaylussacia* (Huckleberry) also bear edible fruits and large-fruited varieties have been raised for commerce. Ornamental plants in this family are dominated by the genera *Erica* (Heath) and *Rhododendron*, both of which contain numerous species, hybrids, and cultivars. Many of these are grown outside in gardens with suitably acid soils, or in glasshouses in the case of the more tender kinds. Other genera of horticultural value include *Arbutus*, *Pernettya*, *Pieris*, *Gaultheria*, and *Kalmia*.

Classification. Some of the genera most often found wild or in cultivation are listed below in the subfamilies to which they are normally assigned:

I. Rhododendroideae (corolla deciduous, fruit usually a capsule)
 Rhododendron (850) chiefly E. Asia and N. America, a few in W. Asia and Europe.
 Kalmia (7) N. America, Cuba.

II. Ericoideae (corolla persistent, fruit usually a capsule)
 Erica (665) chiefly S. Africa, a few on tropical African mountains, Mediterranean region and N. Europe.
 Calluna (1) Europe, Azores, Morocco, W. Asia.

III. Vaccinioideae (corolla deciduous, fruit often a berry)
 Vaccinium (450) N. hemisphere, tropical mountains, S. Africa.
 Gaultheria (150) circumpacific.
 Arctostaphylos (50) chiefly western N. America.
 Gaylussacia (48) N. and S. America.
 Pernettya (20) New Zealand, Tasmania, Mexico to S. America.
 Arbutus (14) western N. America, W. Europe to Mediterranean region.
 Pieris (7) E. Asia, eastern N. America.

Vaccinium oxycoccos, Cranberry

Arbutus unedo, Strawberry Tree

Fig. 42. Two genera with berries.

6.1 Erica herbacea L. (E. carnea L.) Winter Heath

Distribution. Native in coniferous woods and on stony slopes in central Europe, extending locally southwards. It was introduced into Britain by the Earl of Coventry in 1763, and since this date numerous cultivars have been raised. These are often free-flowering and tolerant of a wide range of environmental conditions. In America they are sufficiently hardy to survive north of the Mason-Dixon line.

Vegetative characteristics. A dwarf, evergreen shrub with almost glabrous young stems that have ridges running downwards from the base of the leaves. The linear leaves, in whorls of 4, have their margins rolled back.

Floral formula. K4 C(4) A8 G̲(4)

Flower and inflorescence. The flowers are borne in a leafy, one-sided raceme at the end of the previous year's shoots. In nature, the flowers are pink or rarely white, but cultivars range from pure white to a deep rosy red. The main flowering time is from January to March, but some early-flowering cultivars may bloom in November and late ones in April.

Pollination. Various kinds of bees and butterflies are attracted by the brightly coloured flowers and by the nectar secreted by a disc surrounding the base of the ovary. A visiting insect first comes into contact with the stigma, which receives any pollen that it may be carrying. But when the stamens are disturbed, fresh pollen falls on to its body, to be carried away to the stigma of another flower.

Alternative flowers for study. Alternatives include *E. cinerea* (Bell Heather), which grows on dry moorland, and *E. tetralix* (Cross-leaved Heath) which prefers wet moorland and bogs. Both species flower from July to September. The closely related *Calluna vulgaris* (Heather or Ling) differs from *Erica* in having a petaloid calyx larger than the corolla and a fruit containing only a few seeds.

A. The inflorescence, a one-sided raceme made up of several regular, bisexual flowers.
 Inflorescence: 15 mm

B. A single, pendulous flower, borne on a short, glabrous pedicel that bears 2 bracteoles about half way along. The 4 free sepals are about half as long as the 4-lobed corolla.
 Pedicel: 3 mm Sepal: 3.25 x 1 mm
 Corolla-tube: 5 x 2 mm
 Corolla-lobe: 0.5 x 1 mm

C. A flower shown in an erect position, with calyx and corolla cut lengthwise to reveal the stamens with their long, flattened filaments joined to the edge of the nectar-secreting disc.
 Filament: 4 mm Anther: 1.5 mm
 Style: 5.25 mm

D. The upper portion of a stamen, showing the 2-celled anther, which dehisces by pores to release the pollen grains that are grouped into tetrads.

E. The base of the flower with stamens removed to reveal the superior ovary. Below the ovary is the nectar-secreting disc, and at its apex the lower portion of the single style can be seen.

F. Detail of the capitate stigma at the end of the simple style.
 Stigma: 0.2 mm in diameter

G. L.S. of the ovary and the basal portion of the calyx and corolla (the stamens have been removed).
 Ovary: 0.7 x 0.8 mm Disc: 0.3 x 0.8 mm

H. T.S. of ovary formed from 4 united carpels, each with a single loculus containing a number of ovules on an axile placenta.
 Ovary: 0.8 mm in diameter

A

F

B

D

C

G

E

H

6.2 Rhododendron ponticum L.

Distribution. Native in the Black Sea region and also found wild in S. Spain and Portugal where it grows in woods or by mountain streams on acid soils. It was introduced into Britain in the mid-eighteenth century, and was commonly planted in woods and other shady places. In some areas it has now become a persistent weed.

Vegetative characteristics. A glabrous, erect, evergreen shrub or small tree, sometimes reaching a height of 5 m or more. The entire, elliptic to oblong, alternate leaves have short petioles and are a shining dark green above and paler beneath.

Floral formula. K(5) C(5) A10 G̲(5)

Flower and inflorescence. Several bisexual, bell-shaped, violet-purple flowers, with brown spots in the throat, form clusters at the ends of the branches in May and June.

Pollination. The corolla-colour, scent, and nectar attract numerous pollinating insects, especially bumble-bees. To reach the nectar, the insects must pass over the upward-curving stamens. The mature anthers squeeze out sticky pollen tetrads which adhere to the visiting insects. When these insects visit a flower in the later (female) stage, the pollen is brushed off on to the receptive stigma.

Alternative flowers for study. In areas with calcareous soils, a suitable alternative would be a single-flowered 'Azalea' grown as a pot plant.

A. A large, terminal winter bud containing the inflorescence and enclosed by several deciduous scale-leaves. The foliage buds are much smaller and slenderer.
Inflorescence-bud: 40 x 28 mm
Bud scales: *c.* 30 x 10 mm

B1. A single flower-bud. The bud has a short pedicel with 2 hairy bracts at its base. Below the folded petals is the very small, slightly 5-lobed calyx.
Flower-buds: 20-30 x 10-12 mm
Pedicels: 10-20 mm in bud, lengthening to 30 mm at flowering time.

B2. Details of one of the bracts, showing the point of attachment below the pedicel.
Bract: 8 x 1 mm

C. The flower, showing the slightly irregular 5-lobed corolla.
Flower: 65-80 mm in diameter
Corolla-lobes: *c.* 22 x 20 mm

D. The perianth cut away to reveal the ovary, surrounded by 5 of the 10 curved stamens of differing lengths. The long, simple, curved style, broadening into the stigma, extends well beyond the stamens.
Stamens: 32-44 mm Ovary: 6 x 3 mm
Style + stigma: 45 mm

E. Detail of the superior ovary, with a filament, hairy towards its base, attached to the receptacle. Below the ovary is a lobed, nectar-secreting disc, enclosed by the shallow, slightly lobed calyx.

F. Upper part of stamen. F1: Side view. F2: Front view. The anthers become inverted during development and dehisce by terminal pores.
Anther: 3 mm
Pore: 0.5 mm in diameter
F3: Pollen grains. The pollen grains form tetrads and are linked together by fine, sticky threads.
Pollen grains: 0.05 mm in diameter

G. L.S. of ovary. The simple style arises from a sunken area at the apex. At the base is a portion of the nectar-secreting disc and 2 of the calyx lobes. Numerous ovules are attached to axile placentas.
Ovary: 6 x 2.75 mm

H. T.S. of ovary formed from 5 fused carpels (compare *Erica*). Each carpel has a single loculus containing several ovules on a well-developed placenta.
Ovary (at flowering time): 2.75 mm in diameter

I. The fruit, a capsule, that dehisces by 5 valves to liberate the numerous small, flat seeds. At the centre are the withered placentas and the style. Unlike *Erica*, the corolla is shed well before dehiscence. The empty capsules remain for a long time on the shrub.
Valves of capsule: 12 x 3 mm

Primula veris, Cowslip

Dodecatheon integrifolium, Shooting Star

Soldanella alpina, Alpine Snowbell

Anagallis arvensis, Scarlet Pimpernel

Fig. 43. Flowers of various genera of the Primulaceae.

7 PRIMULACEAE Vent. Primrose family

22 genera and 800 species

Distribution. World-wide, especially N. temperate region.

General characteristics. Herbaceous plants, perennial or sometimes annual, with rhizomes or tubers. Leaves usually entire, mostly opposite or whorled, sometimes all basal, without stipules. Flowers in umbels, panicles, heads or solitary, usually regular, bisexual, with the flower-parts usually in fives, and often heterostylous. Calyx of 5 united sepals, remaining attached. Corolla of 5 usually united petals, or absent (*Glaux*). Stamens 5, opposite the corolla-lobes, occasionally 5 staminodes opposite the calyx-lobes. Anthers 2-celled, facing inwards and usually dehiscing lengthwise. Ovary typically of 5 united carpels, superior or half-inferior, unilocular with free-central placentation, style and stigma simple, ovules usually numerous, arranged spirally or in whorls on the placenta. Fruit a usually 5-valved capsule, dehiscing by teeth at the tip, or a pyxis, in which the top lifts off like a lid (*Anagallis*). Seeds few or many. Embryo small, in fleshy or hard endosperm.

Economic and ornamental plants. The family is best known for decorative plants, particularly species and hybrids of *Primula*, which are commonly cultivated for rock and woodland gardens. The genera *Androsace*, *Dodecatheon*, *Lysimachia*, and *Cyclamen* are also grown. *C. persicum*, the florists' Cyclamen, is a popular houseplant, as are *Primula obconica* and *P. malacoides*, but it should be noted that these species of *Primula* often cause dermatitis.

Classification. The family is divided into 5 tribes:

I. Flowers regular, ovary superior
 1. Primuleae (corolla-lobes overlapping)
 Primula (400) chiefly in hilly regions of the northern hemisphere.
 Androsace (100) northern hemisphere.
 Dionysia (42) mountains of central and W. Asia.
 Dodecatheon (14) chiefly western N. America.
 Soldanella (10) mountains of central and S. Europe.
 2. Cyclamineae (corolla-lobes rolled together, plant tuberous)
 Cyclamen (19) central and S. Europe, N. Africa, Somalia, W. Asia.
 3. Lysimachieae (corolla-lobes rolled together, plant not tuberous)
 Lysimachia (150) world-wide, especially E. Asia and N. America.
 Anagallis (20) mainly W. Europe and Africa.

II. Flowers regular, ovary half-inferior
 4. Samoleae (corolla-lobes overlapping)
 Samolus (15) world-wide, usually in salt-marshes.

III. Flowers irregular
 5. Corideae
 Coris (2) central and western Mediterranean region, Somalia.

Fig. 44. L.S. of Cyclamen flower.

Calyx-lobe
Reflexed corolla-lobe
Stamen
Style
Superior ovary
Receptacle

7.1 **Primula vulgaris** Huds. Primrose

Distribution. A common plant found growing in open woodlands, hedgerows, and meadows throughout the British Isles. Its distribution extends from W. Europe to the Balkans and S. Russia, Asia Minor and N. Africa.

Vegetative Characteristics. A perennial herb with a short, stout rhizome and a rosette of obovate, irregularly-toothed leaves.

Floral formula. K(5) C(5) A5 G̲(5)

Flower and inflorescence. The yellow flowers appear from January to May. They are unusual for the genus in being borne singly, on elongated pedicels. They are of two kinds, 'pin-eyed' and 'thrum-eyed', which differ in the relative positions of the style and stamens. This variation is known as heterostyly.

A. A flower at an early stage, showing the hairy calyx of 5 united sepals closely surrounding the corolla-tube.

B. L.S. of a pin-eyed flower. The superior ovary is situated at the base of the tube, and is crowned by the long style. Above the tube the corolla broadens out into 5 showy lobes.
Calyx-tube: 16 mm Calyx-lobe: 6 mm
Corolla: 30 mm in diameter Anther: 2 mm
Stigma + style: 10.5 mm Ovary: 2 x 2 mm

C. L.S. of corolla-tube of a thrum-eyed flower, showing the shorter style and the stamens at the mouth of the tube.
Stigma + style: 6 mm Ovary: 2 x 2 mm
Anther: 2.5 mm

D. Portions of the corolla-tube of a pin-eyed flower at stamen level. D1: Section showing attachment of stamen. D2: View of anther at dehiscence. The anthers are 2-celled, face inwards, and dehisce lengthwise. Note the constriction in the corolla-tube just below the stamen.

E. T.S. of ovary with free-central placentation. The vascular bundles supplying the ovary are arranged centrally in the placenta.
Ovary: 2 mm in diameter

Pollination. The presence of two different kinds of flower, each kind pollinating the other, helps to ensure cross-pollination. The main pollinating agents appear to be long-tongued bees and long-tongued flies.

Alternative flowers for study. Primula veris (Cowslip) which flowers in April and May, or other species of *Primula* would be suitable. A hybrid between *P. vulgaris* and *P. veris*, with probable admixture of other species, has resulted in the well-known garden plant Polyanthus, which is readily available in flower from March to May.

F. L.S. of ovary, showing how the ovules are attached to the placenta, and the position of the ovary in relation to the corolla and the hairy calyx.

G. The capitate stigma and upper part of style. Stigma: 0.6 x 1 mm

H. The calyx has been partly cut away to show the fruit, a capsule, which even at maturity only just reaches the base of the calyx-lobes. As the capsule ripens the pedicel bends towards the ground, and at dehiscence the 5 valves, each of which is divided into 2 parts at the top, split open allowing the seeds to fall around the plant. The seeds of *P. vulgaris*, unlike most other species of *Primula*, contain an oil in their outer layer and in a specially developed oil-body (elaiosome) which is attractive to ants. These insects quickly carry away the seeds, but after robbing them of their oil, leave them alone to germinate when conditions are suitable. In this way the seeds are dispersed over a wider area. Mature capsule: 11 x 10 mm

A

B

C

D1 D2

E F

G

H

7.1 *Primula vulgaris*, Primrose

Flower of a vigorous hybrid

Fruit, a capsule, opening
by up to 10 teeth

Sprouting tuber

7.2 *Cyclamen persicum*, Cyclamen

7.2 Cyclamen persicum Mill.

Cyclamen

Distribution. Although the name '*persicum*' suggests that this species comes from Persia (Iran), it is in fact native in the eastern Mediterranean region, Algeria and Tunisia, and was first cultivated in France about 1731. Since then plant breeders have developed numerous horticultural strains that are very popular as houseplants.

Vegetative characteristics. A perennial herb with a rounded tuber that becomes somewhat flattened with age. Roots are usually produced from the base of the tuber, and leaves with long petioles arise in a spiral arrangement from the upper surface. The smooth, fleshy leaf-blades are heart-shaped with toothed margins. They are often marbled above and pale green or purplish beneath.

Floral formula. K(5) C(5) A5 G(5)

Flower and inflorescence. The white, mauve or pink flowers arise from the leaf-axils and appear with the mature leaves from December to early May in the wild. When grown as pot-plants they may flower at other times of the year. The flowers are borne singly on purplish brown pedicels that bend over towards

the ground as the fruits develop. In most other species the pedicels coil downwards.

Pollination. Bees are attracted by the sweet scent of the flowers and, in the absence of nectaries, probe the area between the style and stamens for sap. As an insect does so, its head becomes dusted with pollen, which will be carried to the stigma of another flower. Self-pollination is unlikely since the stamens of a flower ripen before the stigma is receptive; also, the style protrudes well beyond the anthers.

Alternative flowers for study. In the interests of conservation, an alternative to the florists' *C. persicum* is not encouraged. However, the species *C. hederifolium* has a wide Mediterranean distribution and is also often grown in gardens. This plant produces a large number of pink flowers from August to October, usually before the leaves. In contrast to *C. persicum*, the leaves of *C. hederifolium* are very variable in size and shape, the flowers are often much smaller, and the pedicels coil downwards in fruit.

A. A flower-bud, showing the twisted petals, and the calyx of 5 united sepals. The calyx-lobes are sparsely glandular, have membranous margins, and a prominent midrib.
 Flower-bud: 20 mm Calyx-lobes: 6 x 5 mm
 Calyx-lobes + receptacle: 6-7 x 7 mm

B. A flower. The flowers are solitary and nodding, and have 5 twisted reflexed petals that are joined shortly at the base. The tip of the style protrudes beyond the corolla-tube.
 Corolla-lobes: 30-35 x 18-25 mm
 Corolla-tube: 7 x 8 mm

C. L.S. of flower, showing the petals leading into the corolla-tube. Part of the calyx can be seen clasping the base of the corolla, also 2 of the 5 stamens that surround the ovary.

D. Base of flower cut open to reveal the stamens that form a close cone around the ovary and lower part of the style. D1 shows 2 of the 5 more or less arrow-shaped stamens with a varying number of papillae on their dorsal surface. D2: One of the stamens has been removed to reveal the superior ovary with its long, tapering style.

E. Detail of stamen. Each stamen is attached to the base of the corolla opposite a lobe. A very short filament supports the anther that dehisces inwards to liberate a large mass of yellow pollen.
 Anther: 6 x 3 mm Filament: 1 mm

F. L.S. of ovary, showing the placenta bearing numerous ovules. On each side of the ovary can be seen part of a stamen, corolla-lobe and calyx-lobe. Note the vascular bundles leading into the ovary from the pedicel. The tapering style is tipped by a minute stigma.
 Ovary + style: 7.75 x 2.5-3 mm

G. T.S. of ovary. Note the free-central placentation and the absence of cross-walls (septa) even though the ovary is composed of 5 fused carpels.
 Ovary: 3 mm in diameter

H. The young fruit surrounded by the persistent calyx. The fruit develops into a capsule that dehisces to about half-way into 4-6 unequal and recurved teeth, allowing the seeds to escape.
 Mature fruit: 14-16 mm in diameter

A

B

H

C

D1

D2

G

F

E

Rubus, Blackberry, Raspberry, etc.
One-seeded drupelets
forming a compound fruit.

Dryas octopetala, Mountain Avens. Group of
achenes on the dry receptacle, and a single
achene with its persistent hairy style.

Filipendula ulmaria, Meadowsweet.
Spirally twisted group of achenes.

Fragaria, Strawberry.
Numerous achenes on
the fleshy receptacle.

Geum, Avens. Group of achenes on the
dry receptacle, and a single achene
with the style persistent as a hooked awn.

Alchemilla, Lady's Mantle. Fruit a single
achene enclosed in the thin, dry receptacle
crowned by the persistent calyx and epicalyx.

Mespilus germanica, Medlar.
Pome crowned by a prominent,
persistent calyx.

Agrimonia, Agrimony. Fruit of
1 or 2 achenes enclosed in the receptacle
which is covered above with hooked spines.

Potentilla, Cinquefoil. Group of achenes
surrounded by the persistent calyx and
epicalyx, and a single achene.

Fig. 45. Fruits of the Rosaceae.

8 ROSACEAE Juss. Rose family

107 genera and 3100 species

Distribution. World-wide.

General characteristics. Trees, shrubs, and herbs, usually perennial. Leaves typically alternate (opposite in *Rhodotypos*), simple or compound, usually with stipules joined to the petiole. Vegetative reproduction often by creeping stems — runners as in *Fragaria* (Strawberry) or suckers as in *Rubus idaeus* (Raspberry). Inflorescence cymose or racemose. Flowers usually bisexual and regular at the end of the stem. Receptacle (hypanthium) generally hollowed. There is often a central protuberance bearing the carpels and in a few cases the carpels are united to the hypanthium. Calyx of 5 free or united sepals, usually overlapping; epicalyx often present. Corolla of 5 free petals, usually overlapping. Stamens in 1 to several whorls of 5 or an indefinite number, bent inwards in bud. Ovary usually with free carpels and superior, sometimes with united carpels or inferior; carpels 1 to several times as many as petals or an indefinite number, 1- to 5-locular, with 1 to several anatropous ovules in each loculus. Placentation various. Fruit various, dry or fleshy, often an aggregate of achenes (*Potentilla*) or drupelets (*Rubus*), or a single drupe (*Prunus*) or pome (*Malus*). In *Fragaria*, the receptacle forms part of the fruit. Endosperm usually absent.

Economic and ornamental plants. The Rosaceae contain a number of genera which provide fruits of economic importance, the main ones being *Malus* (Apple), *Pyrus* (Pear), *Prunus* (Plum, Cherry, Apricot, Peach and Almond), *Rubus* (Blackberry, Raspberry, and Loganberry), *Fragaria* × *ananassa* (Garden Strawberry), and in subtropical countries *Eriobotrya japonica* (Loquat). Others, less widely grown, include *Cydonia oblonga* (Quince) and *Mespilus germanica* (Medlar). *Prunus*, *Malus*, and *Pyrus* are also of considerable horticultural value for their flowers, and in the case of the first two genera numerous cultivars have been raised. Perhaps the most widely cultivated genus for decorative purposes is *Rosa* (Rose) which has been grown in gardens since ancient times and whose named cultivars are now numbered in thousands. Many other genera are represented in parks and gardens, e.g. *Cotoneaster*, *Chaenomeles*, *Pyracantha*, *Kerria*, *Spiraea*, *Crataegus*, *Sorbaria*, *Potentilla*, and *Geum*.

Classification. The family is usually divided into the following 4 subfamilies:

I. Spiraeoideae (ovary superior; carpels 2 or more, each with several to many ovules, ripening as follicles)
 Spiraea (70) N. temperate region.
 Neillia (11) E. Asia.
 Physocarpus (10) N.E. Asia, N. America.
 Sorbaria (7) central and E. Asia.
 Exochorda (4) central Asia, China.

II. Rosoideae (ovary superior; carpels 2 or more, each with 1 ovule, ripening as achenes or drupelets)
 Potentilla (500) chiefly N. temperate region and arctic.
 Rubus (250) world-wide.
 Alchemilla (250) N. temperate region, tropical mountains.
 Rosa (100) N. temperate region, tropical mountains.
 Acaena (100) chiefly S. hemisphere.
 Fragaria (12) Eurasia, N. America, Chile.
 Filipendula (10) N. temperate region.

III. Prunoideae (ovary superior; carpel 1 with 1 ovule, ripening as a drupe)
 Prunus (400) world-wide.

IV. Maloideae (ovary inferior; carpels 2-5, ripening as a pome)
 Crataegus (280) N. temperate region.
 Sorbus (85) N. temperate region.
 Cotoneaster (50) Eurasia, N. Africa.
 Malus (25) N. temperate region.
 Pyrus (20) temperate Eurasia.

8.1 Spiraea × arguta Zab.

Distribution. Spiraea × *arguta* arose as a cross between the hybrid *S.* × *multiflora* (*S. crenata* × *S. hypericifolia*) and another species, probably *S. thunbergii*. It is one of the most popular of spring-flowering shrubs and is often planted in sunny positions in European gardens. It is grown in temperate parts of Australia, and in America, where it is hardy throughout most of New England.

Vegetative characteristics. A bushy, deciduous shrub, 2-3 m high. The slender twigs bear alternate leaves, that are broader and slightly toothed towards the apex.

Floral formula. K5 C5 A15-∞ G5

Flower and inflorescence. Small regular, bisexual, white flowers are borne in clusters along the twigs in April and May.

Pollination. Numerous small insects are attracted by the conspicuous clusters of scented, white flowers as well as by the abundance of nectar secreted from a ring round the concave receptacle. As the flowers are massed together, pollination between flowers in the same cluster can easily occur. Self-pollination may also take place, with or without the aid of insect-visitors.

Alternative flowers for study. In the absence of *S.* × *arguta* there are many species and cultivars to choose from as suitable alternatives. Most of the early-flowering plants have white flowers in umbel-like racemes, while the later-flowering ones often have white, pink or crimson flowers in elongated or flattened panicles. The genera *Aruncus* and *Filipendula* were formerly included in *Spiraea* but are now treated as botanically distinct.

A. Vegetative and flowering shoots. A1: Vegetative shoot in summer showing the alternate leaves, toothed at the apex. A2: Slender twigs with several flower-clusters borne on the previous season's growth. Below each inflorescence are a few entire bracts.

B. Flower-bud, showing the sepals borne on the rim of the concave receptacle (hypanthium).
Bud: 2.5-3 mm in diameter
Sepals: 1.25-1.75 x 2 mm Pedicels: 10-12 mm

C. A regular, bisexual flower. The 5 free petals and 15 or more stamens are joined to the receptacle. The rim of the receptacle has 10 lobes from which nectar is secreted. The central whorl consists of 5 free carpels, the stigmas of which are visible. Five free sepals alternate with the 5 petals.
Flower: 7-10 mm in diameter
Petals: 3.25 x 3 mm

D. L.S. of flower showing the concave receptacle enclosing 2 of the 5 free carpels. The stamens, petals and sepals are attached to the receptacle behind the nectar-secreting rim.

E. Detail of an anther, which is 2-celled, and attached at its back to the filament. The anthers face inwards and dehisce lengthwise.
Filament (at flowering time): 1.7 mm
Anther: 0.5 x 0.5 mm

F. L.S. of ovary. The single loculus contains several pendulous ovules.

G. Detail of the capitate stigma at the end of the simple style.
Stigma: 0.2 mm in diameter.

H. A single, slightly hairy carpel attached to the receptacle which is hairy on its inner surface. The carpel may later develop into a several-seeded follicle though this is often rare in hybrids.
Ovary: 1 x 0.7 mm Style: 1 mm

I. A seed. The ovule lengthens considerably after fertilisation to form a hard, elongated seed.
Seed: 0.5 mm

A2

B

C

D

E

G

A1

F

I

H

8.2 Rosa canina L. Dog Rose

Distribution. Native in most parts of Europe (excluding the extreme north), N. Africa and S.W. Asia, and naturalised in eastern N. America. It is common in hedges, woods, and scrub throughout the British Isles.

Vegetative characteristics. A very variable, deciduous shrub from 1 to 3 m in height, with arching stems armed with curved or hooked prickles. The leaves are pinnate, consisting of 5 or 7 ovate or elliptical, toothed leaflets, and there are prominent stipules at the base of the petioles.

Floral formula. K5 C5 A∞ G̲∞

Flower and inflorescence. The showy, scented, pink or white flowers, either solitary or in groups of 2-4, are borne during June and July on relatively long pedicels.

A. A flower-bud. At the top of the green, urn-shaped hypanthium are relatively long, erect sepals, which protect the folded petals in the bud stage, but become reflexed as the flower develops.

B. Two kinds of sepals. B1: Having an entire margin. B2: Having a lobed margin with gland-tipped divisions. Both kinds as well as intermediate forms may occur in the same flower.

C. L.S. of the regular, bisexual flower with some of the sepals and petals removed to reveal the numerous stamens that are joined to the margin of the prominent disc. The reflexed sepals usually fall away as the fruit matures.
Sepals: 25 x 6.5-7 mm Petals: 35 x 30 mm
Stamens: up to 10 mm

D1. Side view of the disc on the hypanthium, showing the tops of the styles, and the bases of some of the stamens and 3 of the 5 sepals.

D2. The disc viewed from above, with a central opening through which protrude numerous styles with simple stigmas. Round the edge of the disc are the bases of the numerous stamens.
Disc: 6-6.5 mm in diameter

E. Detail of stamen with 2-celled anther.
E1: Young anther before dehiscence. E2: Mature anther, showing the 2 cells dehiscing lengthwise.
Young anther: 1.75 x 1.25 mm
Mature anther: 2 x 2 mm

F. L.S. of the hypanthium that encloses a number of carpels, each one containing a single ovule. The styles bunch together to pass through the

Pollination. Several kinds of insects are attracted by the fragrance of the flowers, their petal colour, and abundant pollen. The visiting insects usually land on the disc within the ring of radiating stamens or on the exposed stigmas at the centre of the flower. The disc has the structure of a nectary but does not appear to secrete any nectar. The stamens curve outwards, well away from the stigmas, so that cross-pollination is favoured.

Alternative flowers for study. Any other single rose would be an acceptable substitute, including the Burnet Rose, *R. pimpinellifolia (R. spinosissima)*, which is found wild on sandy heaths and dunes, and is also cultivated in gardens. It begins to flower about a month earlier than *R. canina*.

narrow opening in the disc, but then spread out to expose the stigmas to insect-visitors.
Hypanthium: *c*. 10 mm Disc: 2 mm

G. A single carpel that ends in a long style, its length varying according to the position of the carpel within the hypanthium. The ovules later develop into hairy achenes enclosed within the fleshy pseudocarp or 'hip' which has developed from the hypanthium.
Carpels: 8-11 mm Ovary: 1 mm in diameter

H. A single achene, which is brownish, wrinkled, and has several long hairs at its apex.
Achenes: 4.75-5 x 3 mm

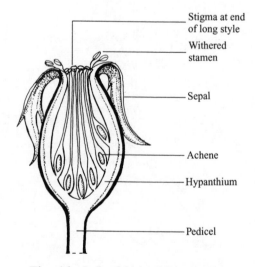

Fig. 46. L.S. of fruit of *Rosa canina*.

A

B2

B1

C

D1

D2

E2

E1

G

H

F

8.2 *Rosa canina*, Dog Rose

8.3 *Prunus spinosa*, Blackthorn, Sloe

8.3 Prunus spinosa L. Blackthorn, Sloe

Distribution. Native throughout most of Europe and commonly found growing in the British Isles in hedges, scrub, and open woodland. It has been introduced into America and has proved hardy in most of New England.

Vegetative characteristics. A deciduous shrub with dark bark, 1-4 m high, often suckering and forming dense thickets. The twigs are softly hairy when young and the older branches are armed with thorns that have developed from numerous short, lateral shoots. The finely toothed leaves are broader towards the apex, narrowing into the petiole. They are dull green above, and usually more or less hairy on the veins beneath.

Floral formula. K5 C5 A20 G̲1

Flower and inflorescence. The solitary white flowers appear from March to May, usually before the leaves. The flower-buds arise from short lateral branches or spurs.

Pollination. Many kinds of insects visit the fragrant, white flowers for their pollen, and for the nectar that is secreted at the base of the hypanthium. The flowers begin to open early in the year when few other flowers are available, and they are particularly conspicuous because they appear before the leaves. The stigma of a flower becomes receptive while the anthers are still closed so cross-pollination must be effected by insects that have previously visited an older flower in the male stage. The stamens subsequently elongate, spread out, and open their anthers. The style also becomes longer but remains receptive, so that should insect-visits fail, self-pollination can take place.

Alternative flowers for study. Any other *Prunus* species or cultivar that has single flowers would be suitable including the edible Plum, *Prunus domestica* ssp. *domestica*.

A. Part of a short lateral branch showing a young flower with the petals more or less erect and the sepals only slightly spreading. At the base of the glabrous or slightly hairy pedicel are several persistent bud scales.

B. The regular, bisexual flower with a portion of the hypanthium cut away to reveal the inner floral parts. The 5 free, spreading petals, reflexed sepals, and numerous stamens are borne on the rim of the hypanthium. At the centre is the ovary ending in a long style.
Corolla: 22 mm in diameter Sepals: 3 x 2 mm
Petals: 8.5-11 x 6.5 mm Hypanthium: 3 x 4 mm
Ovary + style: 11 mm Stamens: 6-9.5 mm

C. Detail of the anther, which is attached at its back to the filament. The anther is 2-celled and dehisces lengthwise.
Anther: 0.6 x 0.5 mm

D. L.S. of hypanthium and ovary. The unilocular ovary contains 2 pendulous ovules (only 1 shown). Surrounding the ovary is the cup-like hypanthium which secretes droplets of nectar. The hypanthium is discarded when the fruit develops.

E. Detail of the capitate stigma with pollen grains. Stigma: 0.7 mm in diameter

F. T.S. of ovary towards the end of flowering time. Two pendulous ovules are situated off centre within the single loculus. After fertilisation one of the ovules develops at the expense of the other so that a single-seeded fruit is formed. Occasionally both ovules develop, producing a 'twin' fruit with the skin and flesh joined and 2 stones at the centre.
Ovary: 0.9 mm in diameter
Loculus: 0.6 mm in diameter

G. L.S. of the mature fruit, a bluish black drupe with a whitish bloom. The fruit is composed of 3 distinct layers — the outer skin (epicarp), a fleshy, green middle layer (mesocarp), and a stony inner layer (endocarp) which encloses the seed. When the seed germinates the endocarp splits along a longitudinal groove.
Fruit: 13 x 11-12 mm

H. The fruit, with part of the skin and fleshy layer cut away to reveal the outer, somewhat pitted surface of the stone.

8.4 Malus × domestica Borkh. Cultivated Apple

Distribution. M. × *domestica* is of hybrid origin and has probably been derived from *M. sylvestris* (Crab Apple), and various other European and Asiatic species. It often escapes from cultivation, sometimes becoming naturalised. More than a thousand cultivars are grown for their fruits in the temperate regions of the world.

Vegetative characteristics. A small to medium-sized deciduous tree with softly hairy young twigs. The dull green, toothed leaves are ovate-elliptical, usually rounded at the base, slightly hairy above and more densely so beneath.

Floral formula. K5 C5 A∞ Ḡ(5)

Flower and inflorescence. The scented, white to pink flowers are borne on spurs on the older wood or at the end of shoots formed in the spring prior to fruiting. The pedicel, hypanthium, and outer surface of the calyx are more or less hairy. Most apple trees flower between late April and the end of May.

Pollination. Bees, flies, midges, and small beetles are attracted to the flowers by their scent, petal-colour, pollen, and by the nectar that is secreted on the inner surface of the hypanthium. The stigmas of a flower become receptive before the anthers shed their pollen, favouring cross-pollination. Domestic apples fall into 2 major genetic categories - diploids and triploids. Diploid apples produce viable pollen but triploids have inferior pollen, so successful pollination is promoted by planting together a mixture of diploid and triploid cultivars with a similar flowering period.

Alternative flowers for study. M. × *domestica* is represented by so many easily available cultivars that there is no real need for an alternative. Comparison between various apple cultivars will reveal numerous variations in flowering and fruiting times, flower size, and especially the size, colour, and shape of the fruit.

A. Fruiting spur, with a group of 4 flowers surrounded by several leaves. The spur has been formed on older wood which bears numerous scars. These scars show where leaves and flowers produced in past years were attached.

B. Young flower with 2 of the 5 petals removed to reveal the numerous stamens surrounding the immature styles. The 5 free sepals are arranged alternately with the petals.
Petals: 18-25 x 10-16 mm
Sepals: 8-12 x 2.75-5 mm
Group of stamens: 6 mm in diameter

C. Two stamens, prior to dehiscence, attached to a portion of the hypanthium. The stamens are usually whorled and are joined to the hypanthium at a point just below the petals. The anthers are versatile, 2-celled, and dehisce lengthwise. Some long, thin hairs are shown at the base of both petals and sepals, but these may be absent in some cultivars.
Stamens: 5-10 mm Anthers: 2.25-3 x 1.75 mm

D. L.S. of a young flower. The 5 sepals and 5 free, clawed petals are joined to the rim of the cup-like hypanthium. The 5 styles are united below and gradually increase in length until at maturity they are as long as or longer than the stamens. The inferior ovary is embedded in the fleshy hypanthium.
Corolla: 40-42 mm in diameter

E1. The 5 styles, showing the numerous thin hairs towards the base. the degree of hairiness varies according to the cultivar concerned.
Styles 10-12 mm

E2. Detail of the upper portion of a style showing the stigma in a receptive state.
Stigma: 0.8 mm

F. T.S. of ovary and hypanthium at flowering time. The ovary consists of 5 united carpels, each carpel containing 2 ovules attached to an axile placenta.
Hypanthium: 3 mm in diameter

G. L.S. of the fruit, a pome, showing the persistent calyx within the apical hollow. The edible part is formed by the development of the ovary wall and the hypanthium. The 'core' consists of the cartilaginous walls of the united carpels which enclose the seeds and protect them from desiccation. The carpels are often 1-seeded owing to abortion of the other seed in the loculus.
Fruit: 55 x 55 mm

Tetragonolobus purpureus, Asparagus Pea. Winged pod, and transverse section showing the 4 wings.

Cytisus scoparius, Broom. Valves of pod coiled after dehiscence.

Cicer arietinum, Chick Pea. Hairy pod, containing 1 or 2 seeds.

Ornithopus sativus, Serradella. Pod constricted between segments, forming a lomentum.

Melilotus officinalis, Ribbed Melilot. Glabrous pod, usually 1-seeded.

Medicago intertexta, Hedgehog Medick. Spirally coiled pod, and detail of spines.

Trigonella caerulea, Blue Fenugreek. Group of 1- or 2-seeded fruits, and a pod abruptly contracted into a beak.

Arachis hypogaea, Groundnut. Reticulate pod which matures underground.

Medicago sativa, Lucerne. Non-spiny, spirally coiled pod, containing 10-20 seeds.

Colutea arborescens, Bladder Senna. Inflated, many-seeded pod, with papery walls.

Fig. 47. Fruits of the Leguminosae.

9 LEGUMINOSAE Juss. (Fabaceae Lindl.)

657 genera and 16400 species — Pea family

Distribution. World-wide.

General characteristics. Trees, shrubs, herbs, water plants, climbers, etc. The roots of most plants have tubercles containing bacteria which enable them to take up much more atmospheric nitrogen. Stems commonly erect, many climbing by leaf- or stem-tendrils or by twining round other plants. Stipules present, sometimes large or represented by spines, e.g. *Acacia*. Leaves usually alternate, often compound. Inflorescence very variable, often a raceme. Flowers usually bisexual, regular or irregular. Calyx of usually 5 more or less united sepals. Corolla of usually 5 free or united petals; in many cases irregular, having a large standard petal (vexillum), 2 wing petals (alae), and 2 petals more or less joined to form a keel (carina). Stamens 10 to many, free or united into a tube. Ovary superior, of usually 1 carpel with 2 to many ovules, anatropous to campylotropous. Fruit extremely variable, typically a pod (legume) opening down both sutures. Seeds with little or no endosperm.

Economic and ornamental plants. A very important family with many genera having edible seeds, e.g. *Pisum, Cicer, Cajanus, Vicia, Phaseolus, Glycine* and *Dolichos*, all various kinds of Peas and Beans, *Lens* (Lentil), and *Arachis* (Groundnut). *Tamarindus indica* (Tamarind) is grown for its fruit in tropical regions. Fodder plants include *Trifolium* spp. (Clover), *Medicago sativa* (Lucerne), and *Onobrychis viciifolia* (Sainfoin). Many genera, especially those in warm regions, provide timber, fibre, dyes, gums, and resins. Oil is expressed from seeds of *Arachis*, and the roots of *Glycyrrhiza* spp. are the source of Liquorice. In temperate regions the plants most commonly grown for ornament include *Lathyrus odoratus* (Sweet Pea), *Lupinus* spp. (Lupins), *Genista* and *Cytisus* (Broom), *Spartium junceum* (Spanish Broom), *Colutea* (Bladder Senna), *Cercis* (Judas-tree), *Laburnum*, and *Wisteria*.

Classification. The family is divided into 3 subfamilies. These are further divided into a large number of tribes (not shown here). Most of the tribes belong to the Papilionoideae.

I. Mimosoideae (flowers regular; petals valvate in bud; leaves often bipinnate)
 Acacia (1200) tropics and subtropics.
 Mimosa (400) mainly tropical and subtropical America.
 Calliandra (200) Madagascar, warm regions of Asia, America.
 Albizia (150) warm regions of Old World, S. America.

II. Caesalpinioideae (flowers usually irregular; lateral petals (wings) covering the standard petal in bud; leaves bipinnate or pinnate, rarely simple)
 Cassia (535) tropical and warm temperate regions.
 Bauhinia (250) tropical regions.
 Caesalpinia (100) tropics and subtropics.
 Gleditsia (14) C. & E. Asia, tropical Africa, N. & S. America.
 Cercis (6) N. temperate region.

III. Papilionoideae (flowers usually irregular; lateral petals enclosed by the standard petal in bud; leaves pinnate, digitate, trifoliolate or simple)
 Astragalus (2000) world-wide, except Australia.
 Indigofera (700) warm regions.
 Crotalaria (600) tropics and subtropics.
 Desmodium (300) tropics and subtropics.
 Oxytropis (300) N. temperate region.
 Trifolium (238) temperate and subtropical regions, except Australia.
 Lupinus (200) America, Mediterranean region.
 Lathyrus (150) N. temperate region, mountains of tropical E. Africa, temperate S. America.
 Vicia (140) N. temperate region, S. America, tropical E. Africa.
 Hedysarum (100) N. temperate region.
 Dalbergia (100) tropics and subtropics, S. Africa.
 Lotus (100) N. temperate region.
 Genista (87) Europe, N. Africa, W. Asia.
 Ononis (75) Canary Islands, Mediterranean region, Europe to C. Asia.
 Medicago (56) temperate Eurasia, Mediterranean region, S. Africa.
 Phaseolus (50) tropical and warm regions of America.
 Cytisus (33) Atlantic Islands, Europe, Mediterranean region.
 Ulex (20) W. Europe, N. Africa.

9.1 Acacia paradoxa DC. (A. armata R.Br.) Kangaroo Thorn

Distribution. Native in temperate Australia, but often grown elsewhere as a glasshouse shrub. In Britain it can only be cultivated outside in the mild S.W. region, preferably against a wall. In the United States it can be grown outdoors in California, Arizona, and other states with similar climates.

Vegetative characteristics. A spreading, evergreen shrub, up to 3 m in height in its native habitat. Its leaves are represented by sharply pointed phyllodes and the stipules at their base are reduced to thin spines.

Floral formula. K(5) C5 A∞ G1

Flower and inflorescence. The flowers appear from March to May. Each inflorescence arises on a peduncle from the axil of a phyllode and consists of a strongly scented, ball-shaped cluster of predominately male flowers together with a few bisexual ones.

Pollination. Various kinds of insects are attracted by the scent of the flowers and their numerous yellow stamens. In the course of its visit the insect's body becomes coated with pollen from the prominent stamens, and this is readily transferred from one inflorescence to another. Small birds may also act as pollinating agents by dispersing the pollen while hunting out the insect visitors.

Alternative flowers for study. In the absence of *A. paradoxa*, *A. dealbata*, the 'Mimosa' of the florist, is readily obtainable. It differs from *A. paradoxa* in that the heads of pale yellow flowers form a panicle, and the leaves are of the normal kind and bipinnate.

A1. Tip of a flowering stem, showing the phyllodes with the central nerve ending a short point. The peduncles bearing the heads of flowers arise from the axils of the phyllodes.

A2. Detail of leaf-axils, showing the axillary buds and the spine-like stipules on the slightly hairy stem. The base of a peduncle can be seen on the right.
Peduncles: *c.* 14 mm
Inflorescences: 6-8 mm in diameter

B. Some of the flowers that form the capitulum. Most of the flowers illustrated are male, but 3 bisexual flowers can be identified by the single style protruding beyond the stamens.

C. A single flower, showing the the 5-lobed calyx surrounding the more or less free petals and the base of 6 of the numerous stamens.
Calyx: 1 mm Corolla: 2 mm

D. L.S. of a bisexual flower. The stamens are inserted round the base of the superior ovary. The ovary is unilocular and contains a row of ovules along the marginal placenta. Each ovule is attached by a funicle which becomes silvery in colour and ends in a cup-shaped aril.
Stamens: 3-3.75 mm Style: 3.75-4 mm
Ovary (at flowering time): 0.9 x 0.25-0.3 mm

E. A stamen. E1: Upper portion of young stamen prior to dehiscence.E2: Upper portion of mature stamen. The anthers are 2-celled and dehisce lengthwise.
Anther: 0.125 mm

F. The slightly hairy ovary, extracted from a bisexual flower, and part of its long style. The position of the ovules is visible through the ovary-wall.

G. The fruit at maturity, a 2- or 3-seeded, straight or slightly curved legume up to 50 mm long. Note that only one of the flowers has developed into a fruit. The remains of the inflorescence can be seen at the base of the legume.
Fruit: 35 x 6 mm

H. The dark brown seed with a well-developed caruncle.
Seed: 7 x 3 mm

A1

A2

C

B

E1

E2

D

F

G

H

9.2 Cercis siliquastrum L. Judas-tree

Distribution. Native in the E. Mediterranean region and W. Asia, but often grown in parks and gardens elsewhere and occasionally naturalised.

Vegetative characteristics. A deciduous tree, up to 10 m in height, with low, spreading branches. The alternate, glabrous leaves are more or less circular in outline, blunt or notched at the apex, and heart-shaped at the base.

Floral formula. K(5) C5 A10 G1

Flower and inflorescence. The purplish rose flowers appear in May, and are borne in clusters of 3-6 on slender pedicels on the old leafless branches or on the trunk itself. Little variation occurs within the species apart from the depth of flower-colour. Cultivars may have flowers varying from white to deep purple.

Pollination. Honey-bees have been observed visiting the flowers in search of nectar which is secreted at their base. The stamens are freely exposed, allowing pollen to be readily transferred to the insect's body and subsequently carried to the stigma of another flower.

Alternative flowers for study. There are relatively few hardy members of the subfamily Caesalpinioideae and, in the absence of *Cercis siliquastrum*, the most readily available alternatives are likely to be species of *Gleditsia*, especially *G. triacanthos* (Honey Locust), a spiny-stemmed tree, occasionally grown in parks and large gardens, which flowers in July. *Caesalpinia gilliesii* and *C. japonica*, which are less hardy, may also be found in flower in July in gardens in the milder parts of the British Isles. Some of the N. American species of *Cassia* can also be cultivated outside in Britain, but other species of *Cassia* and *Caesalpinia* require glasshouse conditions.

A. A single flower with an irregular corolla of 5 petals and a bell-shaped calyx with 5 equal teeth.

B. The petals. B1: The standard.
Claw: 4-5 mm Limb: 6-8 x 5 mm
B2: A wing petal.
Claw: 5 mm Limb: 10 x 6 mm
B3: A keel petal, showing the inner surface.
Claw: 2.5-3.5 mm Limb: 10-11 x 6-8 mm

C. L.S. of flower. Both petals and sepals are overlapping, and the petals vary considerably in width. Five of the 10 free stamens are exposed to reveal their point of insertion. The superior ovary bears a curved style.
Pedicels: 15-17 mm Calyx: 8 mm
Ovary + style (at flowering time) 16 mm

D. The upper portion of a stamen. D1: Ventral view, showing the anther dehiscing lengthwise to shed the pollen grains. D2: Dorsal view, showing the attachment of the anther to the filament.
Filaments: 13-14 mm Anthers: 1.25 mm

E. Lower part of stamen on a portion of the receptacle. Note the tuft of hairs at the base of the inner surface of the filament.
Hairs: 0.25 mm

F. Detail of the upper part of the style, showing pollen grains adhering both to the receptive stigma and to the area just below it.
Stigma: 0.7 x 0.5 mm

G. T.S. of ovary, showing 2 ovules on the marginal placenta. The ovules are arranged in 2 rows, and alternate with each other so as to form 1 vertical rank.
T.S. of ovary: 1 x 0.4 mm

H. The fruit, a legume which is much compressed, and has a narrow wing along the seam (suture) on its ventral side. In autumn the fruits are more or less crimson in colour but they turn brown during the winter and persist for a long time on the tree. Each fruit contains up to 12 black seeds.
Fruits: 50-65 x 10-12 mm
Seeds: 5 x 3-3.5 mm

9.2 *Cercis siliquastrum*, Judas-tree

9.3 *Vicia faba*, Broad Bean

9.3 Vicia faba L. Broad Bean

Distribution. One of the most ancient of all Old World cultivated vegetables and grown as food for cattle, horses, and human beings since prehistoric times. It is considered by some authorities to have been native in S.W. Asia or N. Africa, although no definitely wild plants from these areas have been found. Alternatively, it may have developed under cultivation from the S. European *V. narbonensis*, which it resembles in many characters.

Vegetative characteristics. An erect, entirely glabrous annual with square, usually unbranched, slightly winged stems. The leaves are composed of 2 or 3 pairs of bluish green, fleshy, ovate leaflets and have stipules at their base.

Floral formula. K(5) C5 A(10) G1

Flower and inflorescence. Two to 6 irregular, bisexual, white or occasionally lilac or purple flowers with a black blotch on the wing petals are borne in clusters in the axils of the leaves from May until July.

Pollination. The Papilionoideae are generally well adapted to pollination by bees and *Vicia faba* is no exception. Nectar is secreted by the inner sides of the stamens near their base and accumulates in the staminal tube. Only long-tongued insects therefore can reach the nectar in a legitimate way, although some with shorter tongues have been seen to bite through the base of the flower in order to steal the nectar. Before the flowers are fully grown the anthers dehisce, shedding their pollen on to the tuft of hairs at the end of the style. When an insect alights on an open flower, it presses down the wings and the keel petals, which are connected at the base, and exposes the rigid style. The pollen adhering to the hairs is then transferred to the underside of the insect, and the rubbing of the insect's body causes the stigma to become sticky, and thus receptive to pollen from another flower.

Alternative flowers for study. The Broad Bean is represented by many cultivars which have only minor floristic differences, and it is so widely grown in Britain that an alternative during its flowering period is scarcely necessary. During mild winters and particularly in the early spring *Ulex europaeus* (Gorse) may be found in flower. In the autumn *Vicia sativa* (Common Vetch) and species of *Ononis* (Restharrow) may still be flowering. In the United States, where *V. faba* is not so popular a vegetable, one of the native species of *Vicia*, or one grown as an ornamental plant, could be used as a substitute.

A. L.S. of the irregular flower, showing the tube formed by the stamens enclosed within the 2 united petals known as the keel. One of the black-blotched wing petals can be seen, also a portion of the standard, the largest petal.
Calyx-tube: 7-8 mm Calyx-lobe: 8 mm

B. The petals.
Top - the standard: 24-30 x 12.5-15 mm
Centre - one of the wing petals:
Claw: 10-12 mm Limb: 10-11 x 5-6 mm
Bottom - the 2 united petals forming the keel:
Claw: 10-11 mm Limb: *c.* 7 x 5.5 mm

C. The upper part of the staminal tube, showing the free, curved portions of the 10 stamens. The upturned end of the style can be seen projecting from the staminal tube.
Free part of stamen: *c.* 9 mm

D. Detail of free part of stamen. The stamens have 2-celled anthers which dehisce lengthwise.
Anther: 3.5 x 1.5 mm

E. L.S. of the superior ovary that ends in a brush-like style. The ovary is unilocular, consisting of 1 carpel with marginal placentation. The ovules are arranged alternately to form a single row and are attached by well-developed funicles to the placenta that extends along the ventral side.
Ovary (at flowering time): *c.* 14 mm

F. T.S. of ovary, showing 1 of the ovules joined by its funicle to the single placenta.
T.S. of ovary: 7 x 5 mm
Ovule: 3.25 x 1.5 mm

G. Detail of the apex of the style, showing the tuft of hairs and the stigmatic surface.

H. The young fruit, a pod (legume) which at maturity splits down both sides to release the seeds. The seeds are large and have a leathery seed-coat (testa). The length of the mature pod is variable and depends on the cultivar concerned but is usually from 10 to 20 cm.
Young fruit: 4.5 cm

10.1 *Ricinus communis*, Castor-oil Plant

10 EUPHORBIACEAE Juss. Spurge family

326 genera and 7750 species

Distribution. World-wide, except Arctic region.

General characteristics. Mostly shrubs or trees, some (including all British species) herbaceous, latex often present. Many growing in dry regions, then heather- or cactus-like in habit, or with branches taking on the form and functions of leaves (cladodes). Leaves usually simple, alternate. Stipules usually present, sometimes in the form of hairs, glands, or thorns. Inflorescence usually complex and varied. Flowers usually regular, unisexual, the male and female either on the same plant (monoecious) or on different plants (dioecious). Sepals usually 5, free, sometimes absent. Petals 5, free, often absent. Stamens 1 to many, free or united. Ovary superior, usually of 3 united carpels, 3-locular. Styles 3, usually 2-lobed. Ovules 1 or 2 in each loculus, pendulous, anatropous; placentation axile. Fruit sometimes a drupe, but more commonly a capsule, which splits into 3 carpels; at the same time each carpel opens along its ventral surface to let the seed escape. Seeds usually with abundant oily endosperm; caruncle present in some genera.

Economic and ornamental plants. Most of the Euphorbiaceae are poisonous, but nevertheless the family contains a number of plants of economic importance. Latex from *Hevea brasiliensis* is the main commercial source of rubber. The genus *Manihot* also yields rubber, and from the roots of *M. esculenta* (Manioc or Cassava) the foodstuff tapioca is prepared. The seeds of *Ricinus communis* and *Croton tiglium* contain castor oil and croton oil respectively. *R. communis* is also cultivated in glasshouses or as a summer bedding plant for its handsome foliage. *Euphorbia pulcherrima* (Poinsettia) is grown as a house plant for its striking red bracts, and several hardy species of *Euphorbia* are cultivated in the herbaceous border or on the rock-garden for their coloured glands and foliage. *Codiaeum variegatum*, often incorrectly called 'Croton', is another popular house plant. Numerous cultivars have been produced with leaves of various colours and shapes.

Classification. The majority of the genera listed below are likely to be found in cultivation, though *Mercurialis* and some species of *Euphorbia* may occur in the wild or as garden weeds. The genera mentioned are shown here in the subfamilies accepted in a recent arrangement by Webster.

I. Phyllanthoideae (2 ovules per loculus; no latex)
 Phyllanthus (600) tropical and warm regions.

II. Acalyphoideae (1 ovule per loculus; no latex)
 Acalypha (430) tropical and warm regions.
 Dalechampia (110) warm regions, especially America.
 Mercurialis (8) Mediterranean region and temperate Eurasia.
 Ricinus (1) E. and N.E. Africa to Middle East.

III. Crotonoideae (1 ovule per loculus; latex reddish or yellowish to milky; leaves often lobed or compound)
 Croton (750) tropical and warm regions.
 Jatropha (170) tropical and warm regions, N. America.
 Codiaeum (6) Malesia to Pacific.

IV. Euphorbioideae (1 ovule per loculus; latex whitish; leaves usually simple, rarely lobed)
 Euphorbia (1600) world-wide, especially warm regions.
 Pedilanthus (14) southern N. America to tropical America.

Fruit of 3 united carpels

Stamen

Horned gland

Pedicel bearing stamen

Fig. 48. L. S. of cyathium of *Euphorbia cyparissias*, Cypress Spurge, a species with horned glands.

97

10.1 Ricinus communis L. Castor-oil Plant

Distribution. An annual herb or short-lived perennial, native in E. Africa and the Middle East. It is now cultivated in many tropical and subtropical countries for its seeds which yield an oil used in medicine and in the manufacture of lubricants, paints, and other industrial products. Green, bronze, and reddish-leaved forms are grown as ornamental foliage plants in parks and gardens.

Vegetative characteristics. A variable plant, 1-7 m high, reaching the size of a small tree in the tropics, though usually only 2-3 m when grown as an annual in temperate climates. The stem bears large, alternate, palmately lobed leaves with 5-12 toothed lobes and long, glandular petioles.

Floral formula. Male: K3-5 C0 A∞
Female: K3-5 C0 G̲(3)

Flower and inflorescence. The plant bears clusters of unisexual flowers in terminal panicles during the summer months. The upper part of each inflorescence consists of 1- to 7-flowered cymes of female flowers, while below are 3- to 16-flowered cymes of male flowers.

Pollination. There is a marked tendency to cross-pollination, since most of the upper, female flowers set seed before the lower, male flowers of the same inflorescence reach maturity. The anthers produce abundant pollen and burst explosively. Although the plant is said to be mainly wind-pollinated, at flowering time the glands on the young leaves below the inflorescence exude copious nectar so it is likely that insects are also involved.

Alternative flowers for study. The genus *Ricinus* contains only one species, but the typical green-leaved form and the popular coloured-leaved cultivars are obtainable as seeds or pot plants.

A. A male flower in bud. The membranous perianth is made up of 3-5 free sepals, the petals being absent.
Bud: 8 x 5 mm

B. L.S. of male flower-bud. The perianth protects the numerous, much-branched stamens that are tightly packed together.

C. The male flower. The sepals have spread apart to reveal the cluster of stamens attached to the disc situated at their base. There is no trace of an ovary.
Sepals: 7-8 x 5-6 mm Stamens: *c.* 10 mm

D. Detail of stamen. D1: The much-branched filament with a roundish, pale yellow anther at the end of each branchlet.
D2: Detail of anther, which is 2-celled and dehisces lengthwise to liberate the large amount of pollen.
Anther: 0.3 x 0.4 mm

E. A young female flower with a deciduous, membranous calyx of 3-5 sepals. The superior ovary bears soft, fleshy, bristle-tipped spines. At the top is a short style with 3 deeply divided, papillate stigmatic-branches that become bright purple at maturity.
Sepals: 5 mm Stigmatic-branches: 5 mm

F. L.S. of the superior ovary with 3 loculi, each containing one large, pendulous ovule. Part of the calyx can be seen below the ovary.
Ovary (at flowering time); 10 x 8 mm
Bristle-tipped spine: 3 mm

G. T.S. of the ovary, showing the carpels with well-developed ovules attached to the axile placenta.
Ovary: 8-10 mm in diameter

H. The whole fruit, a spiny capsule which splits into 3 one-seeded portions at maturity. In some forms the seeds are ejected from the carpels, while in others the seeds are retained for some time within the capsule.
Fruit: 20 x 15 mm

I. One of the 3 carpels detached from the axile placenta. The caruncular end of the large seed is visible through the opening caused when the carpel was torn away from the placenta.
Carpel: 20 x 10 mm

J. The smooth, patterned seed, with a prominent caruncle at one end. The seed contains the highly poisonous substance ricin, which is retained in the residue when the oil is extracted.
Seed: 14 x 9 mm

A

D1

D2

E

B

F

G

C

H

I

J

10.2 **Euphorbia helioscopia** L. Sun Spurge

Distribution. Often found as a weed in cultivated ground. It is native throughout the British Isles and other parts of Europe, N. Africa, and Asia. It has been introduced into N. America, and is now found in parts of southern Canada and the eastern United States.

Vegetative characteristics. A glabrous, annual herb with a slender stem and alternate, obovate leaves which are toothed towards the tip.

Floral formula. Male: P0 A1
 Female: P0 G̲(3)

Flower and inflorescence. The cyathium, an inflorescence condensed to simulate a single flower, is characteristic of the genus. Cyathia are borne on branches arising from the leaf-axils. The 5 primary branches or rays have the appearance of an umbel. Subsequent branching is into 2 or 3 raylets. The unisexual flowers appear from May to October, both male and female on the same plant.

Pollination. The 4 oval glands of the cyathium secrete a shallow layer of nectar which is completely exposed. The female flowers mature first, their 3 bilobed stigmas emerging from the involucre and becoming dusted with pollen from other flowers. Later, when the pedicel bearing the ovary lengthens and projects far beyond the involucre, the stamens gradually elongate and dehisce. Pollination is effected exclusively by flies, but beetles, wasps, and bees are occasional visitors.

Alternative flowers for study. Any species of *Euphorbia* would be suitable. Comparison should be made with *Ricinus* and with other genera mentioned in the Classification section.

A. One of the 2 or 3 raylets borne by each of the 5 primary rays, showing further branches bearing raylet leaves and ending in cyathia.
Raylet (1st series): 35 mm
 (2nd series): 10 mm
Peduncle: 3.5 mm
Raylet leaves (decreasing in size upwards):
 36 x 16 mm, 9 x 7 mm, and 7 x 6 mm

B. Close-up of a cyathium. The single female flower, borne on a pedicel arising from the centre of the cyathium, can be clearly seen.
Cyathium: 3 x 1.75 mm Ovary: 1 x 1 mm
Style: 1.25 mm

C. The yellow-green cyathium, formed from 4 united bracts, and with 4 oval glands along the upper edge, is opened up to reveal the individual male flowers, each consisting of a single stamen. The stamens in the centre ripen first, the ripening continuing in a centrifugal manner towards the outer wall.

D. A male flower, showing the single stamen with its short filament attached to a longer pedicel. A clearly visible line denotes the point of attachment. The pedicel of each male flower has a feather-like bract at its base.
Anther-lobe: 0.25 x 0.25 mm
Bract: 1-1.25(-2) mm

E. L.S. of the 3-celled female flower, with a pendulous ovule in each loculus. At the top are the 3 styles, each with a 2-branched stigma.
Ovary: 3 x 3.75 mm

F. T.S. of the ovary, showing the 3 carpels with well-developed ovules; placentation axile.
Ovary: 4 mm in diameter

G. The whole fruit, a capsule, in process of dehiscence. The capsule separates into the 3 carpels, each of which splits into 2 valves leaving the central column standing (see H).
Fruit: 4 x 4 mm

H. Detail of central column of capsule (also known as a columella).

I. A single seed, showing the reticulate surface. At one end is a swelling, the caruncle. A number of species of *Euphorbia* have caruncled seeds, and it has been observed that these are sought by ants, doubtless for their oil content. The structure is thus an elaiosome, and is of importance in aiding seed-dispersal.
Seed: 2 x 1.75 mm

A

B

G

C

D

H

E

F

I

11.1 *Geranium pratense*, Meadow Crane's-bill

11 GERANIACEAE Juss. Geranium family

5 genera and 679 species

Distribution. World-wide.

General characteristics. Mostly herbaceous plants, often hairy. Leaves opposite or alternate, usually lobed or divided, often with stipules. Flowers regular or irregular, bisexual. Calyx of 5 usually free sepals, persistent. Corolla of 5 free petals. Stamens as many or 2 or 3 times as many as petals, more or less united at base, some sometimes reduced to staminodes, e.g. *Erodium* (Stork's-bill); anthers usually versatile, usually shedding their pollen before the stigmas of the same flower are receptive. Ovary superior, of 5 united carpels, with 1 or 2 anatropous ovules in each carpel; placentation axile; styles long, sometimes partially united, but the 5 stigmas free, fruit a schizocarp, the carpels splitting off from a central column (the persistent style), each taking with it a strip of the tissue, called the awn. In most species of *Geranium* the awn suddenly curls upwards, throwing the seed some distance from the plant, while in the remaining species of *Geranium*, and in the other 4 genera, the whole carpel (mericarp) together with the awn is flung off. Except for a few species of *Geranium* where the awn breaks away at the moment of discharge, this structure remains attached. Its stiff apex usually presses against surrounding vegetation, while its lower, coiled portion extends as it takes up moisture, pushing the mericarp into the ground and ensuring that the seed is well placed for germination. Seeds with usually curved embryo and little or no endosperm.

Classification. The family has sometimes been considered to include as many as 14 genera, but a number of these are often placed in other families. The 5 genera most closely related are distinguished as follows:

A. Flowers irregular
 Pelargonium (280) tropical, especially S. Africa.

B. Flowers regular
 Stamens 10, all fertile:
 Geranium (300) world-wide, especially temperate regions.
 Stamens 10, 5 reduced to staminodes:
 Erodium (60) Europe, Mediterranean region to central Asia, temperate Australia, S. America.
 Stamens 15, plants more or less herbaceous:
 Monsonia (25) Africa, S.W. Asia, N.W. India.
 Stamens 15, stems thick and fleshy, armed with spines:
 Sarcocaulon (14) S. Africa.

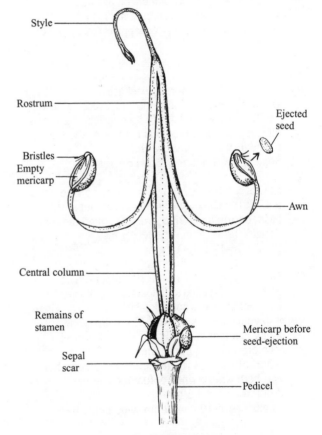

Fig. 49. Seed dispersal in *Geranium pratense*.

103

11.1 Geranium pratense L. Meadow Crane's-bill

Distribution. Native but local throughout most of the British Isles. Widely distributed in Europe, though rare in the Mediterranean region. Also found in N. and central Asia and naturalised in eastern N. America. It has proved to be a useful plant for the herbaceous border.

Vegetative characteristics. A perennial herb with a stout rhizome. The hairy, long-stalked leaves are divided almost to the base into 5-7 coarsely toothed lobes.

Floral formula. K5 C5 A5+5 G̲(5)

Flower and inflorescence. The regular, bisexual flowers are arranged in pairs on peduncles which arise from the axils of the leaves, forming a cymose inflorescence. After flowering the pedicels are reflexed, but they become erect as the fruit develops. The showy, violet-blue flowers appear from June to September or October.

A. Flower-bud on a short pedicel arising from a long axillary peduncle. The pedicel of a second flower can be seen on the left. The 5 overlapping sepals have an awn-like bristle at the tip. At the base of the pedicels are 4 narrow bracts.
Sepals: 8 x 3.5 mm Bracts: 5 mm

B. L.S. of the flower, showing 3 of the 5 free petals. The separate stamens widen out at the base to form a cup-like structure round the ovary.
Sepals: 12 x 7 mm (bristle 3 mm)
Petals: 16-17 x 13-17 mm

C. Dorsal view of one of the 10 stamens. The anthers are 2-celled and dehisce lengthwise. The filament broadens considerably at the base where it extends under the ovary. There are twice as many stamens as petals, the outer whorl of stamens being opposite the petals. The inner whorl of stamens ripens before the outer whorl, and both whorls ripen before the stigmas.
Anthers: 2.5 x 1.5 mm
Filaments: 8-10 mm, 2mm wide at the base

D. All the petals and some of the other flower-parts have been removed to show the ovary at a late stage of development. The style ends in 5 stigmas. When the seed is ripe, the awn suddenly coils up, ejecting the seed through the opening on the inner side of the mericarp.
Ovary + style + stigmas: 14 mm

Pollination. The flowers are visited by bees and other hymenopterous insects which seek out the nectar secreted by the glands at the base of the inner stamens. To prevent self-pollination the anthers ripen first. The 5 stamens in each whorl become erect and shed their pollen. The stigmas then expand and ripen in order to receive pollen from another flower.

Alternative flowers for study. Geranium sylvaticum (Wood Crane's-bill) is a closely related species which differs from *G. pratense* in having petals of more variable length and pedicels which remain erect after flowering. The flowers of most British species are produced in pairs, but in *G. sanguineum* (Bloody Crane's-bill) the flowers are solitary. Annual members of the genus, e.g. *G. robertianum* (Herb Robert), tend to have smaller flowers than the perennials.

E. Detail of the stigmas as they begin to spread. Stigma: 3 mm

F. L.S. of the superior ovary, showing the point of attachment to the style. The ovules are pendulous. The central column (carpophore) eventually acts as an important aid to seed dispersal. On the outer side of the base of each of the 5 inner stamens is a nectar-secreting gland.

G. T.S. of ovary. The ovary is composed of 5 carpels, each containing 2 ovules on an axile placenta. Only 1 ovule of each pair develops into a seed. The ovary increases in width from 2 to 3.5 mm when the fruit is formed.

H. Detail of 2 of the 5 mericarps (central mericarp removed for clarity). The mericarp on the left has been cut open to reveal the young seed. Note the hairs at the apex of the seed, also the small aborted second ovule. At the base of each mericarp is a horny tubercle bearing several hairs.
Mericarp: 5 x 2.5 mm
Young seed: 4 x 2.25 mm

I. Ripened carpel forming a 1-seeded mericarp attached to a portion of the carpophore known as the awn. Part of the carpel-wall has been cut away to reveal the blackish seed. Some of the hairs on the tubercle have turned so as to hold the seed in place until it is ejected.

12.1 *Heracleum sphondylium*, Hogweed, Cow Parsnip

12 UMBELLIFERAE Juss. (Apiaceae Lindl.)

420 genera and 3100 species Carrot family

Distribution. World-wide, chiefly N. temperate region and tropical mountains.

General characteristics. Mostly herbaceous plants with stout stems and hollow internodes. Leaves alternate and sheathing, with their blades much divided pinnately. A few, e.g. *Hydrocotyle* and *Bupleurum*, have entire leaves. Stipules absent. Inflorescence usually a compound umbel (often with an involucre of bracts) formed from partial umbels (each one having an involucre of bracteoles), sometimes a simple umbel as in *Astrantia* and *Hydrocotyle*, or a cymose head as in *Eryngium*. Flowers usually bisexual and regular. Calyx of 5 free or united, usually very small sepals. Corolla of 5 free petals, usually white or yellow, rarely absent. Stamens 5, anthers opening inwards. Ovary inferior, of 2 united carpels, 2-locular, with 1 pendulous, anatropous ovule in each loculus. Fruit a dry schizocarp, which generally splits down a septum (commissure) between the carpels into 2 mericarps, each containing 1 seed. The mericarps are usually held together at first by a thin stalk (carpophore) running up between them. On the outer surface of each mericarp are usually 5 primary ridges, 2 of which (the lateral ridges) are at the edges where the splitting takes place. Between the 5 primary ridges there are sometimes 4 secondary ridges. In the furrows between the ridges there are often oil-canals (vittae). Seed often united to the pericarp, with a small embryo in oily endosperm. The structure of the ripe fruit is of prime importance in the taxonomy of this family.

Economic and ornamental plants. The family contains a number of plants of economic importance, the chief of these being *Daucus carota* (Carrot) and *Pastinaca sativa* (Parsnip). The cultivars which are now grown have much larger and fleshier roots than the wild plants from which they were developed. *Petroselinum crispum* (Parsley), *Foeniculum vulgare* (Fennel), *Anthriscus cerefolium* (Chervil) and *Anethum graveolens* (Dill) are cultivated for their leaves, and *Apium graveolens* (Celery) for its blanched leaf-stalks. The young stems and leaf-stalks of *Angelica archangelica* (Angelica) are crystallised with sugar and used in confectionery. Dill and Fennel are also grown for their fruits, as are *Carum carvi* (Caraway), *Pimpinella anisum* (Anise), *Coriandrum sativum* (Coriander), and *Cuminum cyminum* (Cumin), all of these being used for flavouring. Decorative plants include species of *Eryngium* and *Astrantia*, also *Heracleum mantegazzianum* (Giant Hogweed) the sap of which may cause dermatitis.

Classification. The following classification into 3 subfamilies is based on that of Engler (see References).

I. Hydrocotyloideae (stipules present; fruit with woody endocarp)
 Hydrocotyle (75) world-wide.

II. Saniculoideae (stipules absent; fruit with soft endocarp; style surrounded by a ring-like disc)
 Eryngium (230) tropical and temperate regions, except tropical and S. Africa.
 Astrantia (10) central and S. Europe, W.Asia.

III. Apioideae (stipules absent; fruit with soft endocarp; style on apex of disc)
 Peucedanum (170) temperate Eurasia, tropical and S. Africa.
 Pimpinella (150) chiefly Eurasia and Africa.
 Bupleurum (70) Europe, Asia, Africa, N. America.
 Seseli (65) Europe to central Asia.
 Heracleum (60) N. temperate region, tropical mountains.
 Angelica (50) N. hemisphere, New Zealand.
 Chaerophyllum (35) N. temperate region.
 Oenanthe (30) N. temperate region, mountains of tropical Africa.
 Carum (30) temperate and subtropical regions.
 Ligusticum (25) N. hemisphere.
 Daucus (22) world-wide.
 Apium (20) world-wide.
 Pastinaca (14) temperate Eurasia.
 Anthriscus (12) Eurasia.
 Torilis (12) Canary Islands, Mediterranean region to E. Asia.
 Smyrnium (7) Europe, Mediterranean region.
 Conium (3) temperate Eurasia.
 Petroselinum (3) Europe, Mediterranean region.
 Coriandrum (2) Mediterranean region.
 Foeniculum (1) Europe, Mediterranean region.

12.1 **Heracleum sphondylium** L. Hogweed, Cow Parsnip

Distribution. Native in W. and N. Asia, N. Africa, and most of Europe. It is common in the British Isles, and is found growing by hedges, woods, roadsides, and other grassy places on a wide range of soils. It has been introduced into N. America.

Vegetative characteristics. A stout, erect biennial or short-lived perennial, reaching a height of 2.5 m or more. The hollow, ridged stem may be glabrous or hairy. The leaves are simply pinnate, variously lobed, and hairy on both surfaces; the lower leaves are stalked.

Floral formula. K5 C5 A5 \bar{G}(2)

Flower and inflorescence. The inflorescence consists of an umbel composed of partial umbels, with white or pinkish flowers that appear between June and September.

Pollination. The large, flat-topped inflorescences afford a good landing-ground for many kinds of bees, wasps, beetles, and flies, which visit the flowers for their pollen and nectar. The anthers of the flowers are incurved at first, but then become erect to shed their pollen. Only when they have all dehisced and most of them have fallen do the stigmas mature. In this way self-pollination is avoided.

Alternative flowers for study. An alternative to *H. sphondylium* is scarcely necessary, but comparison should be made with other genera listed in the Classification section.

A. A partial umbel. The outer flowers are larger and open slightly before the smaller, inner flowers.
Compound umbel: 20 cm in diameter
Partial umbels: 25-35 mm in diameter

B. Two flowers taken from different parts of the inflorescence. At the centre of each of the flowers are the 2 united carpels. B1: An outer or marginal flower with petals and stamens of different lengths. B2: One of the inner flowers with petals and stamens of similar lengths. The petals of both kinds of flower are bilobed, with an incurved point between the 2 lobes. The stamens are alternate with the petals.
Petals: 3-10 x 2-3 mm Filaments: 3-10 mm

C. Detail of the upper part of a stamen. The 2-celled anther, facing inwards, and attached at its back to the filament, dehisces lengthwise.
Anther: 1.25 mm

D. A single flower, with 2 petals and 3 stamens removed to reveal the hairy, inferior ovary bearing 2 styles. The stamens and petals are attached to the upper rim of the ovary together with a barely visible calyx represented by 5 minute teeth.

E. L.S. of the inferior ovary, composed of 2 united, unilocular carpels, each of which contains a solitary, pendulous ovule on an axile placenta. The 2 styles have enlarged bases forming a stylopodium.
Ovary: 2.25 x 1.5 mm Stylopodium: 1 x 2 mm
Style + stigma (at early stage): 1 mm

(at late stage): 2.25 mm

F1. T.S. of the ovary, hairy at flowering time.
Ovary: 1.5 x 1.25 mm

F2. T.S. of the young fruit, its ridged pericarp now glabrous. The carpels have grown larger and have become dorsally compressed, and the 2 ovules are developing into flattened seeds. The 2 carpels will eventually split apart down the septum.
T.S. of young fruit: 6 x 2 mm

G. Detail of the receptive stigma, situated on the upper, inner surface of the style.
Stigma: 0.4 x 0.2 mm

H1. The mature, dry fruit before dehiscence, showing the 4 dark lines that indicate the presence of oil-canals (vittae) which are embedded in the fruit-wall between the narrow ribs. At the top can be seen the remains of the 2 styles.
Fruits: 6.5-9 x 6-7 mm

H2. The fruit, which has now split into 2 portions ready for dispersal by the wind. At first the carpels are united by their inner faces, but later they separate along the septum into 2 one-seeded portions known as mericarps. The 2 mericarps are held in position by a divided hair-like stalk, the carpophore, until they are detached by the force of the wind. Note that the mericarp on the right has twisted round.

13.1 *Solanum dulcamara*, Bitter Sweet, Woody Nightshade

13 SOLANACEAE Juss. Nightshade family

90 genera and 2600 species

Distribution. Tropical and temperate regions, especially Central and S. America.

General characteristics. Herbs, shrubs or small trees. Leaves without stipules, entire or variously dissected, usually alternate, but sometimes in pairs in the upper part of the plant. Flowers solitary or in cymes, bisexual, usually regular. Calyx of 5 united sepals, persistent. Corolla of 5 united petals, rarely 2-lipped, usually folded in bud. Stamens 5, or fewer in irregular flowers, alternate with corolla-lobes; anthers often opening by pores. Ovary superior, of 2 united carpels, 2-locular, sometimes with secondary divisions (e.g. *Datura*); ovules 1 to many in each loculus, anatropous or amphitropous on axile, often swollen, placentas; style simple, with entire to 2-lobed stigma. Fruit a berry or capsule. Seed with straight to curved embryo in usually oily endosperm. Flowers conspicuous, insect-pollinated.

Economic and ornamental plants. The principal food plants in the Solanaceae are *Solanum tuberosum* (Potato) and *Lycopersicon esculentum* (Tomato) which are now economically important throughout the world. Numerous cultivars of both of these have been raised. Other widely grown food plants include *Capsicum annuum* (Sweet Pepper), *C. frutescens* (Cayenne Pepper), *Solanum melongena* (Aubergine), *Physalis peruviana* (Cape Gooseberry), and *Cyphomandra betacea* (Tree Tomato). Another plant of great commercial value is *Nicotiana tabacum*, the leaves of which are dried and made into tobacco. Many members of the family are poisonous, and several of these provide drugs used in medicine, e.g. *Atropa belladonna* (Deadly Nightshade), *Hyoscyamus niger* (Henbane), *Datura stramonium* (Thorn-apple), and *Scopolia carniolica*. The family also contains a large number of ornamental plants such as the perennial *Physalis alkekengi* (Chinese Lantern) with its conspicuous, inflated, red calyces, and popular annuals like *Petunia*, *Nicotiana*, *Salpiglossis*, *Schizanthus*, and *Browallia*. The last three are also cultivated as flowering pot plants. *Solanum capsicastrum* and *S. pseudocapsicum*, known as Jerusalem or Winter Cherries, make attractive house plants when bearing their orange, cherry-sized but inedible fruit. Tender genera such as *Brunfelsia*, *Solandra*, *Streptosolen*, *Cestrum*, and the woody species of *Datura* are sometimes grown in heated glasshouses as decorative flowering shrubs.

Classification. The family is divided into 2 subfamilies each with a number of tribes. Here, the genera most likely to be encountered are listed under the subfamily concerned.

I. Solanoideae (Calyx increasing in size with age; corolla regular; fruit a berry; embryo curved)
 Solanum (1400) world-wide.
 Lycium (100) temperate and subtropical regions.
 Physalis (80) world-wide, especially America.
 Hyoscyamus (15) Europe, N. Africa, S.W. and central Asia.
 Capsicum (10) tropical America.
 Datura (8) tropical and warm temperate regions.
 Solandra (8) Mexico to tropical S. America.
 Lycopersicon (7) Pacific S. America.
 Atropa (4) Europe, Mediterranean region to Himalaya.
 Nicandra (1) Peru.

II. Cestroideae (Calyx not increasing in size with age; corolla often irregular; fruit a capsule; embryo straight)
 Cestrum (175) warm regions of America, W. Indies.
 Nicotiana (67) warm regions of America, Australia, Polynesia.
 Brunfelsia (40) tropical America.
 Petunia (35) tropical S. America.
 Fabiana (25) warm temperate S. America.
 Schizanthus (12) Chile.
 Salpiglossis (2) S. Andes.
 Browallia (2) tropical America.
 Streptosolen (1) Peru.

Fig. 50. *Physalis alkekengi*, Chinese Lantern. The papery, orange calyx cut in half to show the fruit inside.

13.1 Solanum dulcamara L. Bitter Sweet, Woody Nightshade

Distribution. Native in Europe, N. Africa and Asia. It is common throughout most of the British Isles and is found in hedges, woodlands, and waste ground. It has been introduced into N. America and has become naturalised near many settlements.

Vegetative characteristics. A scrambling, woody perennial, 30-200 cm, with glabrous to slightly or densely hairy stems and leaves. The leaves are alternate and ovate, sharply pointed at the apex, and truncate, heart-shaped, or arrow-shaped at the base. They may be entire, or with 1-4 lobes or small, stalked, basal pinnae.

Floral formula. K(5) C(5) A5 G̱(2)

Flower and inflorescence. Loose cymes of 10-25 or more regular, bisexual flowers appear from June until September, the bright yellow stamens contrasting strongly with the purple corolla. The flowers have pedicels that are erect during flowering but become recurved in fruit.

Pollination. The flowers are visited mainly by bees for the collection of pollen, since little nectar is produced. The insect clings to the cone of exserted anthers, and by vibrating its wings rapidly causes pollen to be drawn out of the apical pores on to its body. Although the stamens mature at the same time as the stigma is receptive, self-pollination is made less likely because the style protrudes some way beyond the anther-cone.

Alternative flowers for study. Any member of the genus *Solanum* would be a suitable alternative, including *S. tuberosum* (Potato), which has larger, white or purple flowers, and larger fruit, sometimes formed from 3 carpels. *Lycopersicon esculentum* (Tomato) differs mainly in having anthers which dehisce by longitudinal slits instead of apical pores. Examination of the T.S. of a Tomato fruit will show clearly the typical structure of a berry, also the subdivision of a fruit by false septa. Comparison with other genera will reveal differences in corolla-shape, number of stamens, and type of fruit. The genus *Schizanthus* has a more or less 2-lipped corolla, 2 stamens, and 3 staminodes (one of them minute). In the native *Hyoscyamus niger* the fruit is a capsule, the upper portion of which forms a lid that falls off when the seeds are ripe.

A. The regular, bisexual flower attached to an erect pedicel. The 5 lobes of the dark purple corolla are recurved and have 2 green spots at their base. The small calyx, hidden behind the corolla, has 5 rounded lobes.
Calyx: 3 mm in diameter
Calyx-lobes: 0.5-1 mm
Corolla: 18-20 mm in diameter
Corolla-lobes: 9 mm

B. L.S. of the flower, showing the reproductive organs. The superior ovary bears a long style that protrudes beyond the stamens. The stamens are inserted on the corolla-tube and are alternate with the corolla-lobes. The 5 anthers form a cone round the style.
Ovary: 1.8 mm Style + stigma: 6-7.5 mm
Anther: 5 mm

C. Detail of the capitate stigma.
Stigma: 0.13 x 0.375 mm

D. Two of the 5 stamens whose very short filaments are inserted at the top of the corolla-tube. The long anthers lie close to each other but are not strictly united. They dehisce by a pore at the tip to liberate the pollen.

E. L.S. of the superior ovary which is formed by the fusion of 2 carpels. Surrounding the ovary are portions of the calyx, corolla, and stamens.

F. T.S. of the 2-locular ovary, showing the ovules attached to axile placentas. At maturity, the ovary develops into a typical berry with a pericarp divided into 3 distinct layers - the inner, juicy layer (endocarp) with embedded seeds, a fleshy layer (mesocarp), and an outer layer of 'skin' (epicarp). The much larger fruit of the Tomato illustrates even more clearly the general structure.
Ovary: 1.4 mm in diameter

G. The fruit at maturity, a red berry with the persistent calyx at its base. The pedicel that is erect at flowering-time becomes recurved at the fruiting stage.
Fruit: 10 x 9-10 mm

14.1 *Symphytum officinale*, Common Comfrey

14 BORAGINACEAE Juss. Borage family

156 genera and 2500 species

Distribution. Tropical and temperate regions, especially Mediterranean.

General characteristics. Mostly bristly hairy, herbaceous plants, sometimes shrubs or climbers. Leaves usually simple, alternate, and without stipules. Inflorescence usually a spiralled cyme, uncoiling as the flowers open so that newly opened flowers always face in the same direction. Flowers usually bisexual and regular. Calyx of 5 free or basally united sepals, usually overlapping. Corolla of 5 united petals, the lobes usually overlapping or twisted in bud, the tube sometimes with hairy scales between the lobes. Stamens 5, alternate with the corolla-lobes; anthers facing inwards; filaments often with a nectar-bearing disc at the base. Ovary superior, of 2 united carpels, 2-locular, with 2 ovules in each loculus, often becoming 4-locular, with 1 ovule in each loculus, by the growth of a false partition (septum); style entire or lobed, usually arising from the base of the ovary between the lobes. Ovules erect, ascending or horizontal, anatropous. Fruit usually of 4 nutlets, or a drupe. Seed usually without endosperm.

Economic and ornamental plants. A number of species in the following genera are commonly cultivated for decoration, mainly in the herbaceous border: *Heliotropium* (Heliotrope), *Myosotis* (Forget-me-not), *Echium* (Viper's Bugloss), *Anchusa* (Alkanet), *Mertensia* (Virginia Bluebells), *Pulmonaria* (Lungwort), *Borago* (Borage), and *Symphytum* (Comfrey), the last two also sometimes grown as herbs for their medicinal properties.

Classification. Engler (see References) divides the Boraginaceae into 5 subfamilies, but following the narrower concept of the family as given in Willis (see References), the Boraginaceae is restricted here to 2 subfamilies, the second of which is further divided into 5 tribes:

I. Heliotropioideae (style arising from the apex of the ovary; fruit a drupe)
 Heliotropium (250) tropical and temperate regions.

II. Boraginoideae (style arising from the base of the ovary; fruit nutlets)
 A. Flowers regular.
 1. Cynoglosseae (base of style more or less conical; tips of nutlets not projecting above point of attachment)
 Cynoglossum (55) temperate and subtropical regions.
 Omphalodes (28) temperate Eurasia, Mexico.
 2. Eritrichieae (base of style more or less conical; tips of nutlets projecting above point of attachment)
 Cryptantha (100) Pacific America.
 Lappula (50) temperate Eurasia, N. America.
 Eritrichium (30) N. temperate region.
 3. Boragineae (base of style flat or slightly convex; attachment surface of nutlets concave)
 Anchusa (35) Europe, N. America, W. Asia, N. and S. Africa.
 Symphytum (35) Europe and Mediterranean region to Caucasus.
 Alkanna (25-30) S. Europe and Mediterranean region to Iran.
 Pulmonaria (14) temperate Eurasia.
 Borago (3) S. Europe and Mediterranean region.
 4. Lithospermeae (base of style flat or slightly convex; attachment surface of nutlets flat)
 Lithospermum (60) temperate regions.
 Myosotis (50) temperate regions.
 B. Flowers irregular.
 5. Echieae
 Echium (40) Canary Islands, Azores, Europe, W. Asia, N. and S. Africa.

Fig. 51. Flower of *Echium vulgare,* Viper's Bugloss, showing the funnel-shaped corolla-tube; 4 of the 5 stamens are exserted.

14.1 Symphytum officinale L. Common Comfrey

Distribution. This species is found in damp places, especially beside rivers and streams, throughout the British Isles, though it is less common in the north. Its natural distribution extends eastwards across Europe to W. Asia, but it has been introduced into N. America and has become naturalised in eastern parts.

Vegetative characteristics. An erect, stiffly hairy perennial with a thick fleshy root, stout stem, and large ovate-lanceolate leaves.

Floral formula. K(5) C(5) A5 G̲(2)

Flower and inflorescence. Pendulous, cream to purple or pink flowers appear from May to July in spiralled cymes.

Pollination. Only certain kinds of bumble-bees and other insects with long tongues are able to reach, in a legitimate way, the nectar at the base of the corolla. Due to the narrower lower portion of the corolla-tube and the ring of scales between the stamens, even a long-tongued insect is obliged to push its proboscis past the anthers in order to obtain the nectar. The proboscis becomes dusted with pollen which is subsequently transferred to other flowers and in this way cross-pollination is effected. However, some of the shorter-tongued bees make holes in the base of the corolla-tube and steal the nectar without contributing to the pollination of the flower.

Alternative flowers for study. Symphytum × *uplandicum* (Russian Comfrey) is a hybrid between *S. officinale* and *S. asperum* and is the commonest *Symphytum* of roadsides, hedge-banks, and woods, but is usually absent from the waterside habitats occupied by *S. officinale*. Examples of most tribes of the Boraginaceae can be found readily, either as native species or as popular garden plants. A feature of many members of this family is the striking colour-change from pink in bud to blue when the flowers are fully open. *Borago officinalis* (Borage) and species of *Myosotis* (Forget-me-not) are examples of a corolla with a short tube and spreading lobes.

A. L.S. of the regular flower. The tubular, 5-lobed corolla is surrounded by a persistent, hairy calyx with 5 tooth-like lobes. The stamens face inwards and are joined to the corolla-tube. The style arises from below the ovary and protrudes beyond the corolla.
Corolla-tube: 14 mm Style: 15 mm
Sepals: 6-8 mm

B. The calyx, with 2 of its 5 lobes removed to show 2 of the 4 fertilised ovules seated on a nectar-secreting disc on the surface of the receptacle. Each of the 2 carpels is unilocular at first but becomes 2-locular as it matures (see E).
The group of developing fruits is 2 mm across.

C. The detached corolla-tube cut open to show the 5 stamens situated between the corolla-lobes and alternating with the 5 triangular, hairy scales which form the corona.

D. Detail of one stamen, showing the filament joined to the back of the anther near its base. The anther has 2 cells which dehisce lengthwise to liberate the pollen.
Anther: 4.5 mm

E. T.S. of ovary showing the 2 carpels, now divided by a false septum into 4 loculi, and the central style. Each loculus contains one ovule with sub-basal placentation, and at maturity the fruit consists of 4 glossy, black nutlets.

F. L.S. of fruit, showing how 2 of the 4 obliquely orientated seeds are attached to the receptacle. Between them can be seen the basal portion of the style.

G. Detail of the simple stigma at the top of the single style.

15.1 *Lamium album*, White Dead-nettle

15 LABIATAE Juss. (Lamiaceae Lindl.)

224 genera and 5600 species Mint family

Distribution. World-wide, especially the Mediterranean region to C. Asia.

General characteristics. Mainly shrubs or herbs, often with square stems. Leaves usually opposite, simple, without stipules, often hairy, and with epidermal glands secreting volatile oils. Inflorescence racemose or cymose, the cymes at the nodes condensed into a false whorl (verticillaster). Flowers usually bisexual and irregular. Calyx of 5 sepals, united into a bell-shaped or funnel-shaped tube, sometimes 2-lipped, persistent in fruit. Corolla of 5 united petals, usually 2-lipped. Stamens usually 4 (2 long and 2 short), sometimes 2, attached to the corolla-tube, with anthers facing inwards. Ovary superior, of 2 united carpels, on a nectar-secreting disc. Early in development each carpel is divided into 2 loculi, so that the ovary becomes 4-locular as it matures. In many genera the style is gynobasic, arising from the base of the ovary between the 4 portions. Stigma 2-lobed; placentas axile, each with one basal, erect, anatropous ovule. Fruit usually a group of 4 nutlets, each containing one seed, sometimes a drupe. Seed with little or no endosperm.

Economic and ornamental plants. Oils and perfumes are obtained by distillation from *Rosmarinus* (Rosemary), *Lavandula* (Lavender), and *Pogostemon* (Patchouli). Other genera are grown as culinary herbs and are used for flavouring, e.g. *Mentha* spp. (Mint), *Salvia officinalis* (Sage), *Thymus vulgaris* (Thyme), *Satureja montana* and *S. hortensis* (Winter and Summer Savory), *Ocimum basilicum* (Basil), *Hyssopus officinalis* (Hyssop), and *Origanum* spp. (Marjoram and Oregano). Other species of *Salvia* are cultivated for ornament, particularly the red-flowered *S. splendens* and the blue-flowered *S. patens*, also *Monarda didyma* (Bergamot), *Moluccella laevis* (Bells of Ireland – so called from the conspicuous, broadly bell-shaped, green calyces), *Nepeta*, *Physostegia*, *Perovskia*, and *Phlomis*. *Thymus serpyllum* and other aromatic, mat-forming species are sometimes planted between the stones of paved walks. *Solenostemon scutellarioides*, previously known as *Coleus blumei*, is popular as a house plant for its variously coloured leaves, and may also be used for summer bedding.

Classification. The genera of the Labiatae are not readily classified into higher units. Those listed below include some of the larger or more important genera.

A. Style not arising from the base of the ovary; nutlets with lateral attachment.
 Teucrium (100) world-wide, especially Mediterranean region.
 Ajuga (50) Old World temperate regions.
 Prostanthera (50) Australia.
 Rosmarinus (2) Mediterranean region.

B. Style arising from the base of the ovary; nutlets with basal attachment.
 Salvia (900) tropical and temperate regions.
 Hyptis (400) warm regions of America, Africa.
 Thymus (350) temperate Eurasia.
 Scutellaria (300) world-wide.
 Stachys (300) temperate and warm regions.
 Nepeta (250) temperate Eurasia, N. Africa, and mountains of tropical Africa.
 Ocimum (150) tropical and warm temperate regions.
 Phlomis (100) Mediterranean region to China.
 Pogostemon (71) S.E. Asia.
 Solenostemon (60) tropical Africa and Asia.
 Lamium (40) Eurasia, N. Africa.
 Origanum (36) Mediterranean region to E. Asia.
 Satureja (30) N. temperate region.
 Mentha (25) Old World temperate regions.
 Lavandula (20) Atlantic islands and Mediterranean region to India.
 Monarda (12) N. America.
 Physostegia (12) N. America.
 Perovskia (7) N.E. Iran to N.W. India.
 Hyssopus (5) Mediterranean region to C. Asia.
 Moluccella (4) Mediterranean region to N.W. India.

15.1 **Lamium album** L. White Dead-nettle

Distribution. Native in most of Europe, the Himalaya, and Japan, but introduced into Ireland. It is commonly found growing in hedges, waste places and near roadsides. It has been introduced into N. America and has become naturalised in the central and eastern United States and eastern Canada.

Vegetative characteristics. A hairy, perennial herb with a creeping rhizome and a 4-angled, erect stem which reaches a height of between 20 and 60 cm. Its leaves are ovate, coarsely and often doubly toothed, sharply pointed at the apex and heart-shaped at the base.

Floral formula. K(5) C(5) A4 G̲(2)

Flower and inflorescence. The bisexual, irregular, white flowers appear from May onwards, sometimes until December if the weather is mild. They are borne in cymes in the axils of opposite bracts but, having very short pedicels, the flowers appear to be grouped closely round the stem, forming a verticillaster or false whorl characteristic of many of the Labiatae.

Pollination. As the white flowers are in dense groups, they appear more conspicuous to pollinating insects, generally long-tongued bees such as bumble-bees, which seek out the nectar secreted by the disc situated at the base of the ovary. Sometimes short-tongued insects steal the nectar by biting through the base of the corolla-tube. The upper lip of the corolla forms a hood, protecting the stamens and stigma from possible damage, while the lower lip acts as a landing-stage for pollinating insects. On arrival, the insect grasps the 2 lateral lobes of the flower, causing the hooded upper lip to descend and press first the style and then the stamens on to its back. As the style projects beyond the stamens it picks up the pollen brought in from another flower before pollen from its own flower is deposited on the insect's back. Cross-pollination is usually achieved although, as the stamens and style mature at the same time, there appears to be no barrier to self-pollination.

Alternative flowers for study. Any other species of *Lamium* would be suitable, including the equally common *L. purpureum* (Red Dead-nettle).

A. L.S. of the bisexual, irregular flower with its 2-lipped corolla surrounded at the base by the tubular, 5-lobed calyx. The hooded upper lip, formed from 2 united lobes, protects the stamens and style which lie close to its inner surface. The 4 stamens (only 2 of which are shown) are didynamous, the outer pair being longer then the inner. The stamens are attached to the lower part of the corolla-tube and alternate with the corolla-lobes, the upper lip counting as a single lobe. The long style, that is 2-branched at the apex, arises from the base of the ovary.
Calyx-tube: 5-6 mm Corolla-tube: 7-8 mm
Upper lip of corolla: 12.5 mm Lower lip: 6 mm

B. The persistent, tubular calyx, with 5 more or less equal lobes or teeth. At the base of the tube can be seen the fruit consisting of 4 nutlets.
Calyx-lobes: 4.5 mm

C. Ventral and lateral views of the hairy anthers, which are 2-celled, face inwards, and dehisce lengthwise.
Long stamen: 6 mm Short stamen: 4 mm
Anther: 1.5 mm

D. L.S. of one of the lobes of the 4-lobed, superior ovary. The single ovule is erect, and is attached to a basal placenta derived from an axile type.
Ovary: 1.5 mm

E. T.S. of the ovary formed from 2 united carpels. At a later stage of development there is a further division of the ovary resulting in 4 separate loculi, each containing a single ovule.

F. T.S. of the mature fruit showing the division into 4 separate portions, each portion becoming a one-seeded nutlet.
Fruit: 1.75 mm in diameter

G. Basal portion of style with one of the ovary-lobes and part of the nectar-secreting disc that is situated below the ovary. The style is gynobasic and arises from the base of the ovary, between the 4 lobes.

H. The 4-lobed ovary at maturity showing the gynobasic style. The disc is more developed on the front side of the ovary.
Nutlet: 2 x 1.25 mm

I. Apical portion of the 2-branched style. The receptive stigmas are at the tips of the branches.
Style (from base to fork): 20 mm
Style-branches: 1.5 mm

15.2 Ajuga reptans L. Bugle

Distribution. Native in most parts of Europe, also S.W. Asia, E. Algeria, and Tunisia. It is common in damp woods, meadows and pastures throughout the British Isles.

Vegetative characteristics. A perennial herb reaching a height of 10-30 cm, with a short rhizome and long, leafy, rooting stolons. The basal leaves form a rosette, and the square stems are often hairy on two opposite sides.

Floral formula. K(5) C(5) A4 G̲(2)

Flower and inflorescence. The bisexual, irregular blue (rarely pink or white) flowers appear from May until July. The inflorescence consists of usually 6-flowered whorls (verticillasters) that are borne in the axils of leaf-like bracts often tinged with blue. The lower flowers always open first.

Pollination. Bees and other insects are attracted to the flowers and seek out the nectar that is secreted by the nectary at the base of the ovary. They land on the 3-lobed lower lip and follow the nectar-guides that lead to the centre of the flower. The stamens project well beyond the corolla-tube and ripen before the stigma. Pollen from the anthers is transferred to the head of the visiting insect. At this stage, the stigma stands well above the stamens. Later on, the style bends down, bringing the stigma near to the entrance of the flower. It is then in a receptive state and in a position to receive pollen brought by insects from other flowers.

Alternative flowers for study. Of the 3 species native in the British Isles, *A. reptans* is the most common. There are also a number of cultivars that show variation in the colour of their leaves and flowers. Comparison with the closely related genus *Teucrium* will reveal a difference in the arrangement of the corolla-lobes, and the absence of the ring of hairs within the corolla-tube.

A. The whole plant, showing the racemose inflorescence, and a stolon growing out from amongst the basal rosette of leaves.

B. A flower-bud, about to open. The hairy calyx has lobes of varying lengths and is irregular at the base in order to enclose the swollen lower portion of the corolla-tube. Some of the hairs have been removed for clarity.
Calyx: 6 mm
Calyx-lobes: up to 4 mm
Corolla-tube: 10 x 3 mm

C. Corolla with the upper lip removed. The base of the corolla-tube is enlarged to accommodate the prominent nectary. The stamens are didynamous, i.e. the 2 pairs are of different lengths. The style ending in a forked stigma protrudes well beyond the mouth of the corolla-tube. The lower, 3-lobed lip is marked with nectar-guides.
Corolla: 15 mm
Corolla-tube: 8 mm
Lower lip of corolla: 7 mm
Corolla-tube: 2.5-3 mm in diameter

D. Part of the corolla-tube showing the point of attachment of the 4 stamens. The upper portion of each stamen is free while the lower part is joined to the corolla-tube down to its base.

Free portion of long stamen: 4.5-5 mm
Free portion of short stamen: 3.75 mm

E. Detail of anthers. E1: before dehiscence, and E2: after dehiscence, showing some of the pollen grains. The anthers are attached at their back to the filaments and dehisce lengthwise. Glandular hairs are often present on the filament.
Anther: 0.6 x 0.2 mm

F. Detail of the base of the corolla-tube. The 2-celled ovary contains 4 nutlets, and at its base is a prominent nectary. Part of the style can be seen arising from the top of the ovary. A ring of hairs is situated about 2 mm above the base of the corolla-tube.
Style: 13 mm
Group of nutlets: 1 x 1 mm
Nectary: 1 x 1.5 mm

G. T.S. of ovary showing the 2 united carpels, each with 2 loculi. There is a single ovule in each loculus. At the base is the large nectary.
T.S. of ovary and nectary: 1.75 x 1 mm

L.S. of flower

A. Flower at male stage B. Flower at female stage

Fig. 52. Pollination in *Salvia pratensis*, Meadow Clary.

In common with many other plants, the reproductive organs of *Salvia pratensis* mature at different times to promote cross-pollination. In any flower of this species, the anthers become ripe before the stigma is receptive.

Attracted by the blue flowers, long-tongued bees land on the lower lip of the corolla. In order to reach the nectar secreted at the base of the ovary, a bee must push past a barrier formed by the lower parts of the stamens. As it does so, the upper parts of the stamens, bearing the fertile anther-lobes, swing downwards dusting its back with pollen.

This lever mechanism, apparent at the male stage, is formed by the connectives of the 2 stamens.

The portions bearing the fertile anther-lobes elongate to form the upper part of the lever, while the lower parts, consisting of the sterile anther-lobes, fuse together to create a plate-like barrier. The pivot point of the lever is situated where the 2 parts of the connective join the filament.

Subsequently, the long, curved style lengthens even further and bends downwards so that its forked tip, bearing the stigma, blocks the mouth of the corolla. When a pollen-laden bee, searching for nectar, visits a flower at the female stage, it finds its way barred by the forked tip of the style. As it brushes past, it leaves pollen on the stigma that is now in a receptive state.

Lamium album,
White Dead-nettle

Lamium amplexicaule,
Henbit Dead-nettle

Galeopsis speciosa,
Large-flowered Hemp-nettle

Clinopodium vulgare,
Wild Basil

Melittis melissophyllum,
Bastard Balm

Nepeta cataria,
Cat-mint

Lycopus europaeus,
Gypsywort

Ajuga reptans,
Bugle

Salvia verbenaca,
Wild Clary

Fig. 53. Flowers of various genera of the Labiatae.

Scrophularia nodosa,
Common Figwort

Mimulus guttatus,
Monkey flower

Pedicularis sylvatica,
Lousewort

Euphrasia nemorosa,
Eyebright

Linaria vulgaris,
Common Toadflax

Digitalis purpurea,
Foxglove

Melampyrum pratense,
Common Cow-wheat

Cymbalaria muralis,
Ivy-leafed Toadflax

Calceolaria
hybrid

Penstemon
hybrid

Paulownia
tomentosa

Veronica chamaedrys,
Germander Speedwell

Fig. 54. Flowers of various genera of the Scrophulariaceae.

16 SCROPHULARIACEAE Juss. Foxglove family

222 genera and 4500 species

Distribution. World-wide, especially temperate regions.

General characteristics. Mostly herbs or undershrubs, a few shrubs or trees (e.g. *Paulownia*), some climbers. Several genera, including *Euphrasia* and *Pedicularis*, are semi-parasitic. Leaves simple or rarely compound, alternate, opposite or whorled, without stipules. Inflorescence racemose or cymose, or flowers solitary in axils (e.g. *Linaria*). Bracts and bracteoles usually present. In *Castilleja*, the upper leaves and bracts are brightly coloured. Flowers bisexual, usually irregular, sometimes nearly regular (e.g. *Verbascum*). Calyx of usually 5 united sepals. Corolla of usually 5 united petals, often 2-lipped. Stamens usually 4 (sometimes 2), the posterior stamen sometimes represented by a staminode (e.g. *Scrophularia* and *Penstemon*). *Verbascum* and its allies have 5 stamens. *Veronica* has 4 sepals (the posterior one absent), 4 petals (the posterior pair united) forming a corolla with a short tube and spreading lobes, and 2 stamens. Other variations in flower structure occur within the family. Nectar-secreting disc present below the ovary. Ovary superior, of 2 united carpels, 2-locular with axile placentation; ovules usually numerous, less commonly few (e.g. *Veronica*), anatropous; style simple or bilobed. Fruit usually a capsule, sometimes a berry or drupe, surrounded by the persistent calyx; seeds usually numerous, small, with endosperm; embryo straight or slightly curved. Flowers usually adapted to insect-pollination.

Economic and ornamental plants. The only economic plant of importance in this family is *Digitalis purpurea* (Foxglove) whose leaves are the source of drugs. It is also an attractive garden plant. Other genera of horticultural value for the herbaceous border include *Antirrhinum, Penstemon, Calceolaria, Mimulus, Nemesia, Verbascum,* and *Veronica.* The genus *Hebe*, comprising evergreen shrubs, is widely planted, particularly in coastal areas. A number of hybrids and cultivars with white, or various shades of pink and purple, flowers are now available. A few species of the tree genus *Paulownia* are in cultivation, especially *P. tomentosa*, which has fragrant, pale violet flowers.

Classification. The following classification into 3 subfamilies is based on that of Wettstein (see References).

I. Verbascoideae (The 2 posterior lobes of the corolla cover the lateral lobes in bud; leaves usually alternate; stamens often 5)
 Verbascum (360) temperate Eurasia.

II. Scrophularioideae (The 2 posterior lobes of the corolla cover the lateral lobes in bud; at least the lower leaves opposite; usually 4 stamens, the 5th absent or represented by a staminode)
 Calceolaria (300) Mexico to S. America.
 Penstemon (250) chiefly N. America.
 Scrophularia (200) chiefly temperate Eurasia.
 Mimulus (150) chiefly America, some in Asia and S. Africa.
 Selago (150) tropical and S. Africa.
 Sutera (130) tropical and S. Africa, Canary Islands.
 Linaria (100) N. temperate region, especially Mediterranean region.
 Nemesia (65) tropical and S. Africa.
 Diascia (50) S. Africa.
 Antirrhinum (42) Pacific N. America, W. Mediterranean region.
 Zaluzianskya (35) chiefly S. Africa.
 Paulownia (6) E. Asia.
 Alonsoa (6) tropical America.

III. Rhinanthoideae (The 2 posterior lobes of the corolla are covered in bud by one or both of the lateral lobes)
 Euphrasia (450) chiefly N. temperate region.
 Pedicularis (350) N. hemisphere.
 Veronica (250) chiefly N. temperate region.
 Castilleja (200) chiefly N. and S. America.
 Hebe (75) Australasia, temperate S. America.
 Rhinanthus (45) temperate N. hemisphere.
 Melampyrum (35) N. temperate region.
 Digitalis (19) Europe to C. Asia, Mediterranean region.

16.1 **Verbascum thapsus** L. Aaron's Rod, Great Mullein

Distribution. Native in Asia as far south as the Caucasus and the Himalaya, and eastwards to W. China, also found wild in most of Europe, though absent from the extreme north and much of the Balkan peninsula. In the British Isles it is widespread and locally frequent on dry, grassy banks and waste places. It is naturalised in N. America.

Vegetative characteristics. An erect biennial, 30-200 cm in height, with stem and leaves densely clothed with soft, whitish hairs. The basal leaves form a rosette and the stem-leaves are alternate. All the leaves are elliptical to obovate-oblong and entire, and the base of the stem-leaves is continued down as wings on the round stem.

Floral formula. K(5) C(5) A5 G(2)

Flower and inflorescence. The inflorescence usually consists of a dense, terminal raceme, and axillary racemes are rarely present. The numerous yellow, slightly irregular flowers appear from June to August on very short pedicels arising from the axils of softly hairy bracts.

Pollination. The conspicuous yellow flowers forming the long raceme remain open even in wet weather, allowing bees and other insects to collect pollen without difficulty. No nectar is produced. The pollen-collecting insect alights on the lowest corolla-lobe, which is larger then the others. The filaments are clothed with long hairs and these enable the insect to maintain a firm hold while it collects pollen, at the same time transferring pollen from another flower on to the receptive stigma. In the absence of cross-pollination, self-pollination can take place.

Alternative flowers for study. There are a number of minor floral differences between species of *Verbascum*, but any other member of the genus may be used if *V. thapsus* is not available. In *V. nigrum* (Dark Mullein) the hairs on the filaments are purple, and in the *V. daenzeri* group from the Balkans only 4 fertile stamens are present.

A. The flower in bud. The calyx is covered in soft, whitish hairs and has 5 more or less equal lobes. Hairs are also present on the outer surface of the 5-lobed corolla.
Flower-bud: 6 mm Calyx-lobes: *c.* 3 x 1.5 mm

B. L.S. of the bisexual, slightly irregular flower. The lowest lobe of the corolla is slightly larger than the others. Three of the 5 stamens are shown, the filaments of the 3 upper (2 shown) are always hairy, while those of the 2 lower (1 shown) may be with or without hairs. The superior ovary bears a long style with a capitate stigma. Numerous ovules are attached to the axile placentas.
Stamens: *c.* 7 mm Ovary: 2 x 1.5-2 mm
Style + stigma: 10 mm

C. Detail of stamen. C1: The anther from an upper stamen. The anther is kidney-shaped and is attached at its centre to the filament. The lower stamens may be similar to the upper ones, or the anther may lie against and appear to recline on the apex of the filament. C2: At dehiscence, the anther splits open lengthwise to shed its pollen. Some pollen grains can be seen adhering to the anther-wall.
Anthers: 5 x 0.75-1 mm

D. The hairy, superior ovary with all the surrounding flower-parts removed except for 3 of the 5 calyx-lobes.

E. Detail of the capitate stigma, showing the papillae.

F. T.S. of ovary. The ovary is formed from 2 fused carpels and has 2 loculi and axile placentation.
Ovary (at flowering time): 1.5-2 mm in diameter

G. The fruit, a 2-celled capsule containing numerous seeds. The persistent calyx can be seen at its base.
Fruit: 8 x 8 mm

H. One of the many brownish furrowed seeds.
Seeds: 0.75-0.9 x 0.5 mm

A

B

C2

D

E

H

F

C1

G

16.1 *Verbascum thapsus*, Aaron's Rod, Great Mullein

16.2 *Antirrhinum majus*, Snapdragon

16.2 Antirrhinum majus L. Snapdragon

Distribution. Native of S.W. Europe eastwards to Sicily, and often grown in gardens from which it sometimes escapes and becomes naturalised, mainly on old walls.

Vegetative characteristics. A perennial, sometimes treated as an annual in gardens, reaching between 30 and 80 cm in height in the British Isles. The erect or straggling herbaceous stem becomes woody near the base. The entire leaves, opposite below and alternate above, are lanceolate or linear-lanceolate, narrowing at the base into the petiole.

Floral formula. K(5) C(5) A4 G(2)

Flower and inflorescence. The flowers form terminal racemes from July to September. The corolla of wild plants is usually pink or purple with a yellow palate, but cultivars may range in colour from pure white to yellow, orange, pink, red, and deep purple.

A. The bisexual, irregular flower is borne on a short pedicel which arises from the axil of a bract. The calyx is deeply 5-lobed and considerably shorter than the corolla.
Corolla: 35-37 mm Corolla-tube: *c*. 20 x 12 mm
Calyx-lobes: 6-7 x 5-6 mm

B. L.S. of the flower. The corolla is prominently 2-lipped, the upper lip is 2-lobed and the lower lip 3-lobed. Stiff, capitate hairs, arranged in interrupted lines within the corolla-tube, act as nectar-guides. A small pouch (appearing wrinkled due to sectioning) is situated at the side of the superior ovary. The ovary ends in a long, curved style with a capitate stigma. Two of the 4 stamens which arise from under the base of the ovary can be seen.
Ovary (at flowering time): 3.5 mm
Style + stigma: 17-18 mm

C. L.S. of the flower, cut horizontally to show the 2-lobed upper lip behind the 4 stamens. The stamens are in 2 pairs of different lengths. The anthers of the 2 longer stamens are fully developed while those of the shorter stamens are less mature.
Longer stamens: 22 and 24 mm
Shorter stamens: 14 and 18 mm

D. Detail of the upper portion of a stamen prior to dehiscence. D1: Front view. D2: Back view. The anthers are 2-celled, face inwards, and dehisce lengthwise.
Anther: 3 x 3 mm

Pollination. In *Antirrhinum* and certain related genera the lower lip of the corolla has a palate with 2 swellings which fit accurately into 2 depressions in the upper lip. The 2 lips can only be prised open by large, strong insects such as bumble-bees. The bee alights on the lower lip, and forces its way into the corolla in search of nectar which collects round the bases of the filaments. As the bee probes for nectar, its back rubs against the anthers and pollen is transferred. When the insect leaves, it carries with it the pollen which is removed by the receptive stigma of another flower.

Alternative flowers for study. A. *majus* and its numerous cultivars are commonly available and no alternative is necessary. Comparison should be made with genera in the other subfamilies of the Scrophulariaceae.

E. L.S. of the lower part of the flower. The ovary protrudes into the basal pouch. Nectar is secreted at the base of the ovary but does not usually flow into the pouch. The basal part of 2 of the filaments can be seen bearing stiff, capitate hairs. Hairs are also present on the ovary and the style.
Ovary: 3.5 x 4 mm

F. T.S. of ovary, composed of 2 united carpels, showing the 2 loculi each with numerous ovules attached to the well-developed axile placenta.
T.S. of ovary: 4 x 2 mm

G. Detail of the capitate stigma.
Stigma: 1.5 mm in diameter

H. The fruit after dehiscence, showing the persistent calyx and the withered style. The 2-celled capsule opens by 3 small pores. The cell nearer the stem is longer, narrower above, and opens by a single apical pore. The other cell is shorter, wider above, and opens by 2 apical pores. The pores are covered by small teeth that curl back at dehiscence to liberate the numerous small seeds.
Fruit: 13-14 x 7-8 mm

16.3 Veronica persica Poir. Common Field Speedwell

Distribution. Native in W. Asia, and first recorded in Britain in 1825, it is now a very common weed of cultivated ground throughout most of Europe. It has been introduced into N. America and has become naturalised in many places.

Vegetative characteristics. An annual herb, with procumbent, softly hairy stems bearing leaves that are mostly alternate though the lower ones may be opposite. The leaves are broadly ovate, and have a toothed margin, a truncate base, and a short petiole.

Floral formula. K(4) C(4) A2 G(2)

Flower and inflorescence. The flowers are borne on long pedicels that arise from the leaf-axils. The corolla is bright blue, with the lower lobe often sonewhat paler or even white. Flowering can occur throughout the year, though many more flowers are produced during the warmer months.

Pollination. Many kinds of insects are attracted by the bright blue corolla and the nectar secreted by a disc situated below the ovary. If the weather is warm the flowers open in the early morning, and the filaments spread well away from the stigma to prevent self-pollination. Pollinating insects, directed by the guide-marks on the corolla-lobes, insert their proboscis into the short corolla-tube in order to seek out the nectar which is protected from rain by the short hairs near its base. On arriving at a flower the insect clings to the stamens and becomes dusted with pollen. At the same time its underside, which may be covered with pollen from another flower, brushes against the stigma. In this way cross-pollination is achieved. On damp mornings the stamens do not spread out so far, so the anthers may then deposit their pollen on to the stigma, effecting self-pollination.

Alternative flowers for study. Any other species of *Veronica*, or any member of the closely related genus *Hebe*, would be a suitable alternative. In *V. spicata* (Spiked Speedwell) and related garden plants, the inflorescence is a terminal raceme, and in *V. chamaedrys* (Germander Speedwell) the racemes are axillary.

A. Portion of a flowering shoot showing the procumbent stem, rooting at the nodes and supporting solitary flowers borne on long pedicels arising from the leaf-axils.

B. The flower in bud. The calyx has a short tube and is deeply divided above into 4 ciliate lobes, 2 of which are smaller than the others.
Calyx-lobes (at flowering time): *c.* 4 x 2.5 mm

C. View of the flower from above. The corolla has a very short tube with 4 unequal lobes. Dark lines converge towards the centre of the flower and act as nectar-guides.
Corolla: 7.5 mm in diameter
Upper lobe: 3 x 4.5 mm Lower lobe: 3 x 1.5 mm
Lateral lobes: 2 x 4 mm

D. L.S. of the corolla, showing the ring of nectar-protecting hairs at the base of the tube. One of the 2 stamens is shown. The anthers are 2-celled, face inwards and dehisce lengthwise. The filaments narrow considerably towards the base.
Filament: 3 mm Anther: 1 mm

E. Back view of one of the 2 stamens, showing the anther attached to a more or less swollen filament.

F. L.S. of the ovary after flowering time with the calyx below it. The ovary is hairy, and is flattened at right angles to the septum.
Ovary: 2.5 x 3.75 mm Style + stigma: 2 mm

G. Detail of the upper portion of the style and the stigmatic papillae.
Stigma: 0.2 x 0.3 mm

H. T.S. of ovary. The ovary is formed by the fusion of 2 carpels and has 2 loculi, each with several ovules on an axile placenta.

I. Fruit with persistent style and calyx. The 2-lobed capsule is laterally compressed, and dehisces along the midrib to liberate the broadly elliptical seeds that are concave on one face.
Fruit: 5 x 8 mm Longer calyx-lobes: 9 x 4 mm
Shorter calyx-lobes: 7 x 4 mm

17.1 *Campanula rotundifolia*, Harebell

17 CAMPANULACEAE Juss. Bellflower family

87 genera and 1950 species

Distribution. World-wide.

General characteristics. Mostly perennial herbs,
usually containing latex, with simple, alternate
leaves. Stipules absent. Inflorescence usually
racemose, sometimes cymose (e.g. *Canarina*).
Flowers usually bisexual, regular or irregular. Calyx
usually 5-lobed, united with the ovary. Corolla of
usually 5 united petals, valvate, often blue or violet.
Stamens 5; anthers facing inwards, sometimes united.
Ovary inferior, rarely superior, of usually 2-5 united
carpels, usually multilocular, with axile placentas
bearing numerous anatropous ovules; style simple,
stigmas as many as carpels. Fruit a capsule, dehiscing
in various ways in different genera, or a berry. Seeds
with abundant fleshy endosperm.

Economic and ornamental plants. Several species of
Lobelia have been used in herbal remedies and one of
these, *L. inflata* (Indian Tobacco), has been employed
in Western medicine as the source of a drug. The
family is horticulturally important mainly because of
the genus *Campanula*, which is represented in gardens
by a large number of species, hybrids and cultivars.
Other genera containing ornamental plamts include
Lobelia, *Platycodon*, *Codonopsis*, and *Phyteuma*.

Classification. The classification into 3 subfamilies
is based on that of Schönland (see References). The
subfamily Lobelioideae is sometimes treated as a
separate family, the Lobeliaceae.

I. Campanuloideae (flowers usually regular; anthers
 usually free)
 Campanula (300) N. temperate region,
 especially Mediterranean region, and tropical
 mountains.
 Wahlenbergia (150) chiefly S. temperate
 region.
 Phyteuma (40) Europe, Mediterranean region,
 Asia.
 Jasione (20) Europe, Mediterranean region,
 W. Asia.
 Platycodon (1) N.E. Asia.

II. Cyphioideae (flowers irregular; filaments
 sometimes united, anthers free)
 Cyphia (50) Africa, especially S. Africa.
 Nemacladus (10) S.W. United States, Mexico.

III. Lobelioideae (flowers usually irregular; anthers
 united)
 Lobelia (365) mostly tropics and subtropics.
 Centropogon (230) tropical America.
 Siphocampylus (215) tropical America.

Fig. 55. Flower-head and single
flower of *Jasione laevis*.

Fig. 56. *Lobelia cardinalis*, Cardinal Flower.
The flower is irregular and has united anthers.

137

17.1 Campanula rotundifolia L. Harebell (Bluebell in Scotland)

Distribution. Native throughout N. temperate regions including the British Isles. It is locally common in dry, grassy places and on fixed dunes, often on poor, shallow soils.

Vegetative characteristics. A slender perennial with ascending to erect stems reaching a height of 15-40(-60) cm. The basal leaves, which are sometimes present at flowering time, are ovate to roundish, have somewhat toothed or lobed margins, and long stalks. The lower stem-leaves are lanceolate, remotely serrate, and stalked, and the upper stem-leaves are linear, entire, and without stalks.

Floral formula. K(5) C(5) A5 $\bar{\text{G}}$(3)

Flower and inflorescence. The blue (rarely white) regular, bisexual, nodding, bell-shaped flowers appear solitarily or in a lax, branched panicle from July until September.

Pollination. The usual pollinating-agents are bees, which are sufficiently agile to reach the nectar secreted by a disc at the base of the style and hidden by the expanded bases of the stamens. While the flowers are still in bud, the anthers shed their pollen on to the hairy, upper part of the style. At flowering time, the filaments have already shrivelled and the style has extended. As a bee, foraging for nectar, crawls into the corolla, it picks up pollen from the style. A few days later, the 3 style-branches separate, exposing the stigmatic surface. An insect now visiting the flower is likely to transfer pollen from another flower on to the mature stigmas and so effect cross-pollination. In the absence of insect visits, self-pollination can take place, since, before withering, the style-branches curve back and allow the stigmas to touch any pollen remaining on the style.

Alternative flowers for study. Any other species of *Campanula* would be a suitable alternative, including the popular garden plant *C. medium* (Canterbury Bell) which flowers in May and June, and is one of the group with a 5-branched style. *Jasione* and many species of *Phyteuma* bear a head of flowers, and *Platycodon* has a 5-celled capsule opening by apical valves.

A. A flower-bud, showing the corolla surrounded by the 5 linear calyx-lobes. The ribbed calyx-tube is joined to the inferior ovary.
Bud: 12 x 3 mm

B. The regular, bisexual flower, shown as erect though it is nodding on the plant. The corolla has 5 lobes that are shorter than the tube.
Corolla: 14 x 13 mm Calyx-lobe: 5 mm

C. L.S. of flower, showing the stamens and style. The 5 stamens alternate with the 5 corolla-lobes, and the expanded bases of the filaments form a dome over the nectar-secreting disc at the base of the style. The anthers have shrivelled and the 3 branches of the style have begun to separate in order to expose the stigmatic surface.

D. A complete stamen. The 2-celled anther faces inwards and dehisces lengthwise. It is attached by its base to a short filament which is considerably enlarged at the base and edged with hairs.

E. The reproductive organs in the male and female stages.
E1: The style and 2 of the 5 stamens as they appear inside the flower-bud. At this stage, the style is only as long as the stamens, and the anthers are able to deposit their pollen on to the hairy, upper part of the style.
Filament: 2.5 mm Anther: 5 mm
Style + stigma: 7.5 mm
E2: The style and 2 of the stamens at the beginning of flowering. The style has lengthened, extending its pollen-covered portion well beyond the stamens. The stamens have begun to wither after shedding their pollen. (See C for next stage of development).
Style + stigma: 10 mm

F. L.S. of the inferior ovary, showing the ovules on axile placentas. Above the ovary is the nectar-secreting disc, protected by the expanded bases of the filaments.
Loculus of ovary: 2 x 0.5 mm

G. T.S. of ovary formed from 3 united carpels, showing the 3 loculi with numerous ovules in each loculus.
Ovary: 1.75 mm in diameter

H. The fruit before dehiscence. The withered corolla and calyx enclose the capsule which opens by basal pores.
Capsule: 5 x 2.5 mm

A B C

E1

F D

G E2 H

Lonicera periclymenum, Honeysuckle, in fruit;
each fruit is a red, few-seeded berry

Weigela floribunda. The inflorescence is a corymb
of red, funnel-shaped flowers

Linnaea borealis, Twinflower, a trailing plant
with hairy stems and pink, bell-shaped flowers

Lonicera involucrata, Twinberry. The yellow flowers,
sometimes tinged with red, are borne in pairs with
reddish or purple bracts on axillary peduncles

Fig. 57. Flowers and fruits of the Caprifoliaceae.

18 CAPRIFOLIACEAE Juss. Honeysuckle family

16 genera and 365 species

Distribution. Mostly N. temperate region and tropical mountains.

General characteristics. Shrubs (sometimes climbing) or small trees, rarely herbs. Leaves opposite, usually simple (pinnate in *Sambucus*), usually entire. Stipules small or absent. Flowers bisexual, regular or irregular, usually in a cymose inflorescence. Calyx of 4 or 5 free or united sepals. Corolla of 4 or 5 united petals, the lobes overlapping or valvate, sometimes 2-lipped. Stamens 4 or 5, joined to the corolla-tube and alternate with the lobes; anthers usually facing inwards (outwards in *Sambucus*). Ovary inferior, of usually 3-5 united carpels, with 1 to many pendulous ovules in each loculus; placentation axile; style simple with capitate stigma. Fruit usually a fleshy berry or drupe, sometimes a capsule; seeds with abundant fleshy endosperm.

Economic and ornamental plants. Some species of *Viburnum* bear edible fruit. If *Sambucus* is included in the family, this genus may also be mentioned as being of local economic importance, the flowers and berries of *S. nigra* (Elder) being used in making wine and the pith in microscopic work. Many genera in the Caprifoliaceae are of horticultural value as ornamental flowering shrubs, including *Viburnum*, *Lonicera* (Honeysuckle), *Symphoricarpos* (Snowberry), *Weigela*, *Diervilla*, *Kolkwitzia*, *Leycesteria*, and *Abelia*.

Classification. The genus *Sambucus* is usually included in the Caprifoliaceae, but is sometimes placed in a separate family, the Sambucaceae, on account of the pinnate leaves, the anthers facing outwards, and certain other features. The following are some of the principal genera:

Lonicera (180) N. hemisphere.
Viburnum (150) temperate regions and subtropics.
Abelia (30) Himalaya to E. Asia, Mexico.
Sambucus (20) temperate regions and subtropics.
Symphoricarpos (17) N. America, China.
Weigela (10) E. Asia.
Leycesteria (6) W. Himalaya to S.W. China.
Diervilla (3) eastern N. America.

Fig. 58. *Lonicera involucrata,* Twinberry. The 2-flowered inflorescence and L.S. of the saccate corolla.

Fig 59. *Lonicera sempervirens,* Trumpet Honeysuckle, and *L. japonica,* Japanese Honeysuckle, species with more elongated corollas.

18.1 Viburnum opulus L. Guelder-rose

Distribution. Native in Europe, N. and W. Asia, and N. Africa. It is common throughout most of the British Isles in woods, hedgerows, and scrub, particularly on damp soils. It has been introduced and has become naturalised in parts of N. America.

Vegetative characteristics. A deciduous shrub, 2-4 m in height, with opposite leaves that are glabrous above and slightly hairy beneath. The leaf-blade has 3-5 irregularly toothed lobes and usually has an attractive red tint in autumn.

Floral formula. K(5) C(5) A5 \overline{G}(3)

Flower and inflorescence. The inflorescence consists of a rather flat-topped cluster of regular, white flowers that appear in June and July. They are followed in September and October by groups of bright red fruits.

Pollination. The small, central flowers are for normal reproductive purposes, while the sterile, outer ones, several times larger, render the cluster more conspicuous to pollinating insects. Both pollen and nectar are easily available to short-tongued insects, so that bees, flies, and beetles visit the flowers and, in landing upon the inflorescence, effect cross-pollination. Because of the proximity of the receptive stigma to the mature anthers, self-pollination may also occur.

Alternative flowers for study. V. lantana (Wayfaring-tree) is common on calcareous soils in the south of Britain, and V. tinus (Laurustinus), an evergreen species from the Mediterranean region, is frequently cultivated. Most Viburnums have entirely bisexual flowers, but a few (like *V.opulus*) have some sterile flowers, while in V. opulus 'Roseum' (Snowball-tree), all the flowers are sterile. *Weigela* and *Diervilla* are examples of genera with capsular fruits.

A. The inflorescence, a flat-topped cluster of small, fertile flowers surrounded by much larger, sterile flowers.
 Sterile flowers: 17-18 mm in diameter
 Fertile flowers: 4-5 mm in diameter

B. Part of the inflorescence, showing one flower in bud and another fully open. The bases of 2 other flowers can be seen. There are often small bracts at the bases of the pedicels. The corolla is regular, with a bell-shaped tube and 5 lobes that are overlapping in bud. The 5 stamens are well exserted from the tube.
 Corolla-lobes of fertile flowers:
 1.25 x 1.25-2.25 mm

C. The 5-lobed corolla, showing the 5 stamens joined to the base of the tube and alternating with the lobes. The 2-celled anthers face inwards and dehisce lengthwise.
 Stamens: 5 mm

D. A versatile anther prior to dehiscence.
 Anther: 1 x 1 mm

E. The corolla and stamens have been removed from the flower to reveal the 3-lobed, almost sessile stigma. The 5 calyx-teeth are arranged round the top of the inferior ovary, and at its base is a portion of the pedicel.
 Stigma: 0.75 x 0.6 mm
 Calyx-teeth: 0.3 x 0.2 mm

F. L.S. of the ovary and stigma. F1: Two of the 3 pendulous ovules on axile placentas. All 3 ovules are only visible in the early stages of ovary development. Each ovule is in a separate loculus. F2: At a later stage, showing the one remaining ovule. The ovary is composed of 3 fused carpels, but in the course of development 2 of the carpels are normally suppressed.
 Ovary: 2 x 1.25 mm
 Ovule in F1: 0.25 x 0.125 mm
 Ovule in F2: 0.7 x 0.125 mm

G. T.S. of the ovary at the later stage, shown by the single loculus, surrounded now by a mass of cellular tissue (stippled in drawing). A small space at the centre denotes the remains of an aborted carpel. The 5 small circles inside the margin are vascular bundles. The ovule has been moved to a horizontal position in order to show how it is attached to the placenta.
 T.S. of ovary: 1.5 mm

H. One of the roundish, red fruits attached to its stalk (the others have been removed for clarity). The fruit is a drupe and contains a single seed or 'stone'. The scars on the stalk indicate where pedicels have fallen away.
 Fruit: 10 x 7 mm

A

B

C

D

E

F1

F2

G

H

18.2 Lonicera periclymenum L. Honeysuckle

Distribution. Native in most of Europe and in Morocco, and common throughout the British Isles in woods, hedges and scrub. As it is one of the most fragrant species, several cultivated varieties are grown in gardens in Europe and the United States.

Vegetative characteristics. A deciduous, scrambling, woody climber reaching 6 m. The opposite, entire leaves are ovate or elliptical, dark green above and glaucous beneath.

Floral formula. K(5) C(5) A5 \overline{G}(3)

Flower and inflorescence. The irregular, bisexual, scented flowers are without pedicels, and are borne in heads at the ends of shoots from June until September. The corolla is usually creamy white to yellow inside, while the outside may have a reddish tinge. There is, however, considerable variation in the colour, particularly after pollination has taken place, when the flowers turn a somewhat darker shade.

Pollination. Long-tongued, night-flying moths are attracted by the creamy white corolla, the presence of nectar, and the strong scent. The secretion of nectar by glands in the basal half of the corolla-tube and the emission of scent continue for several days, but gradually diminish as the flower ages. On the first night of opening the flower is in the male phase, the stamens projecting horizontally from the corolla-tube, while the style is bent down towards the lower lip to avoid self-pollination. The following day the flower enters the female phase. The corolla-tube now curves slightly downwards and has become yellowish in colour, while the stamens have moved closer to the lower lip and the style has risen to a point directly opposite the entrance to the flower. The alternate prominence of stamens and style promotes cross-pollination.

Alternative flowers for study. *L. periclymenum* and its cultivars are readily obtainable. The two other native species, *L. caprifolium* (Perfoliate Honeysuckle) and *L. xylosteum* (Fly Honeysuckle), flower in May and June. Many honeysuckles are upright shrubs including *L. fragrantissima*, which bears small, sweetly scented flowers from December until March, and *L. involucrata*, whose pairs of flowers appear in May. The latter has broad, reddish bracts, which become darker with age.

A. An inflorescence. Some of the irregular, bisexual flowers are still in bud while others are fully open.

B. L.S. of the corolla, with the inferior ovary and the 5-toothed calyx left intact. The corolla is 2-lipped, the upper lip being 4-lobed and the lower one entire. The well-exserted stamens are joined to the corolla-tube just inside the mouth, and the long style, ending in a 3-lobed stigma, projects even beyond the stamens. Part of the long, narrow nectary (shown as a wavy line) can be seen on the lower inner surface of the corolla-tube. Both corolla and calyx have small glandular hairs on their outer surface.
Calyx-teeth: 1.5 x 1 mm
Corolla-tube: 20-25 mm
Upper lip: 10-20 x 12 mm Lobes: 4-8 x 3 mm
Lower lip: 17-20 x 4 mm Style + stigma: 50 mm

C. Detail of a portion of the corolla when opened out. The upper part of the drawing shows how 4 of the 5 stamens are attached to the corolla-tube (the central portion is the base of the lower lip). The lower part shows the longitudinal ridge in the basal half of the corolla-tube that bears the nectar-secreting glands.
Nectary: 7-11 mm

D. The upper part of a stamen. The versatile, 2-celled anthers face inwards and dehisce lengthwise.
Filament: 20 mm Anther: 5 mm

E. L.S. of the inferior ovary, with the glandular calyx and base of the corolla-tube. Within the tube is the style with a small swelling at its base. The ovary is composed of 3 fused carpels, each with a single loculus containing 2 pendulous ovules attached to axile placentas. On one side of the ovary is part of a glandular bract.
Ovary: 3 mm

F. T.S. of the ovary, showing the 3 loculi with 2 ovules in each loculus. The attachment of the ovules to the placentas is not visible here.
T.S. of ovary: 2.75 x 2.5 mm

G. The young fruit, a red, few-seeded berry crowned by the persistent, glandular calyx. At its base are small, glandular bracts.

A

D

C

E

F

B

G

Portion of a capitulum with both disc- and ray-florets

Disc-floret of *Doronicum*

Ray-floret of *Doronicum*

Tubular floret of *Centaurea*

Fig. 60. Florets in the Compositae.

19 COMPOSITAE Giseke (**Asteraceae** Dumort.)

1317 genera and 21000 species Daisy family

Distribution. World-wide.

General characteristics. Mainly herbaceous plants, rarely shrubs or trees, usually with a taproot, sometimes with tubers (e.g. *Dahlia*). Leaves usually alternate, sometimes opposite, rarely whorled, simple or compound, frequently in basal rosettes, without stipules. Oil passages usually present. Latex sometimes present. Inflorescence racemose, the flowers (florets) arranged in heads (capitula), the heads often grouped into larger inflorescences. Each head consists of a group of florets, borne on a receptacle, and surrounded by usually green bracts. Calyx usually represented by a pappus, sometimes a slightly 5-lobed rim on the top of the ovary. Corolla of 5 united petals, usually consisting of a long or short tube and a variously shaped limb. Stamens 5, with short filaments, alternate with the corolla-lobes; anthers facing inwards, forming a tube round the style. Ovary inferior, of 2 united carpels, unilocular, containing one erect, anatropous ovule with basal placentation; style simple, branching into 2 stigmas. Fruit a cypsela, often crowned by the persistent pappus which enlarges after fertilisation to aid distribution. Endosperm absent; embryo straight.

Economic and ornamental plants. This large family contains relatively few plants of economic value. The seeds of *Helianthus annuus* (Sunflower) provide an edible oil, and the tubers of *H. tuberosus* (Jerusalem Artichoke) are grown for food, as are the roots of *Tragopogon porrifolius* (Salsify) and *Scorzonera hispanica* (Scorzonera), the leaf-stalks of *Cynara cardunculus* (Cardoon), the young flower-heads of *C. scolymus* (Globe Artichoke), and the leaves of *Lactuca sativa* (Lettuce) and *Cichorium endivia* (Endive). Many genera listed in the Classification section include well-known decorative plants. In addition, *Callistephus* (China Aster), *Dendranthema* hybrids (Garden Chrysanthemums), *Echinops* (Globe Thistle), *Onopordon* (Scotch Thistle), *Solidago* (Golden Rod), *Coreopsis*, *Cosmos*, *Gaillardia*, and *Rudbeckia* are often found in gardens. *Senecio* × *hybridus*, the florist's Cineraria, is a popular house plant.

Classification. The Compositae are well-marked in their characters, and are unlikely to be confused with any other family. Two subfamilies are generally recognised, each divided into a number of tribes.

I. Lactucoideae (florets usually of one kind; disc-florets usually with long, narrow corolla-lobes, usually purple, pinkish or white, sometimes yellow; anthers dorsifixed; latex sometimes present)
Vernonia (1000) tropical and warm regions.
Centaurea (450) mainly Mediterranean region.
Crepis (200) chiefly N. hemisphere.
Cirsium (200) N. temperate region.
Lactuca (100) world-wide, mostly N. temperate region.
Carduus (91) Eurasia, N. and E. Africa.
Taraxacum (60) mainly N. temperate region.
Mutisia (59) S. America.
Arctotis (50) tropical and S. Africa.
Hypochoeris (50) Eurasia, N. Africa, S.America.
Eupatorium (38) mostly America, a few in Eurasia.
Gerbera (35) tropical and S.Africa to S.E.Asia.
Gazania (16) tropical and S. Africa.
Lapsana (9) temperate Eurasia.
Cichorium (8) Europe, Mediterranean region, temperate Asia, Ethiopia.

II. Asteroideae (florets usually of different kinds; disc-florets usually with short, broad corolla-lobes, usually yellow; anthers basifixed; latex absent)
Senecio (1500) world-wide.
Helichrysum (500) warm regions of Old World.
Artemisia (300) N. temperate region, S. Africa, S. America.
Aster (250) America, Eurasia, Africa.
Erigeron (200) world-wide, mainly N.America.
Ligularia (180) temperate Eurasia.
Inula (90) Eurasia, N. Africa.
Helianthus (67) America.
Tagetes (50) warm regions of America.
Helenium (40) America.
Ursinia (40) S. Africa.
Doronicum (35) temperate Eurasia, N. Africa.
Dahlia (28) Mexico to Colombia.
Zinnia (22) S. United States to Argentina.
Calendula (20) Mediterranean region.
Tussilago (15) N. temperate region.

19.1 Taraxacum officinale Weber Dandelion

Distribution. Common throughout the N. hemisphere and abundant in the British Isles.

Vegetative characteristics. A perennial herb with a basal rosette of coarsely toothed leaves, from which rises a hollow scape bearing a solitary capitulum.

Floral formula. K pappus C(5) A(5) \overline{G}(2)

Flower and inflorescence. The solitary inflorescence consists of a head of yellow florets surrounded by 2 rows of green bracts, the inner erect, the outer erect, spreading or even reflexed. The florets are all bisexual and ligulate. The flowering period is from March until October.

Pollination. Various kinds of insects are attracted by the conspicuous head of florets and by the nectar secreted round the base of the style. Since the florets are massed together, many can be visited by a single insect in a short time, and as the abundant nectar collects and rises in the corolla-tube, it becomes accessible to all but the shortest-tongued insects. The anthers form a cylinder, and when they dehisce the pollen is trapped by the hairs on the outer surface of the stigmas at the top of the style. As the style lengthens, the stigmas project sufficiently for insects to gather the pollen. At a later stage the stigmas separate, exposing the receptive areas to pollen from other flowers. In this way cross-pollination may take place. However, like most other species of *Taraxacum*, and many species of *Hieracium* (Hawkweed), *T. officinale* is usually apomictic. The term apomixis covers the various methods of reproduction that have been developed as substitutes for the sexual production of seeds. The effect is that the progeny of apomictic plants are genetically identical to their parents, and those genera concerned are commonly divided into large numbers of 'microspecies' composed of almost identical individuals.

Alternative flowers for study. The Dandelion is so common that an alternative is hardly necessary. Other closely related plants include *Lapsana communis* (Nipplewort) and *Hypochoeris radicata* (Cat's-ear) which are widely distributed throughout the British Isles. *Cichorium intybus* (Chicory), which is often cultivated, could also be used. All these flower during the summer and early autumn.

A. Capitulum at an early stage, showing the involucre of glabrous bracts surrounding the numerous protruding florets. The inner, larger bracts are erect and the outer, smaller bracts are reflexed.
Outer bracts: 7-10 x 3-5 mm
Inner bracts: 14-21 x 2-4 mm

B. L.S. of a capitulum. The florets are all bisexual and irregular.
Receptacle: 10 mm in diameter
Inflorescence: 4.5 cm in diameter

C. A single floret, showing the inferior ovary, and the corolla surrounded by the hairy pappus. The corolla is extended on one side into a long strap-like structure and is termed ligulate.
Floret: 21 mm Pappus: 7 mm Beak: 1 mm
Cypsela: 2 mm

D. The upper part of the corolla has been rolled back to reveal more clearly the reproductive organs. The 5 short filaments are borne on the corolla-tube and are separate, but the anthers are united and form a tube round the style. The style extends far beyond the stamens, and branches into 2 stigmas covered with small protuberances (papillae) which help the pollen to adhere.
Filament: 1.5 mm Anther: 4 mm
Style: 12 mm Stigma: 2 mm

E. L.S. of young ovary, showing the single, basal ovule.

F. T.S. of ovary. Both ovary and ovule are somewhat compressed.
Ovary (at flowering time): 0.9 x 0.5 mm

G. A mature fruit with the pappus attached to the cypsela by a long beak (absent in the genus *Doronicum*). This type of structure, common within the Compositae, provides a very efficient method of fruit dispersal.
Cypsela: 3.5 mm
Beak: 12 mm
Pappus: 7 mm

H. Enlarged view of the cypsela, showing the longitudinal ribs.

19.1 *Taraxacum officinale*, Dandelion

19.2 *Doronicum* × *excelsum*

19.2 Doronicum × excelsum (N. E. Br.) Stace

Distribution. Commonly grown in gardens for its early flowering, this plant is thought to be a hybrid between three species, *D. pardalianches* , *D. plantagineum* and *D. columnae*.

Vegetative characteristics. A perennial herb with numerous, stout, underground stolons, tuberous at their tips, which produce rosettes of long-stalked, ovate-elliptical leaves with prominent, curving lateral veins.

Floral formula. K pappus C(5) A(5) \overline{G}(2)

Flower and inflorescence. The inflorescence is usually solitary on a long stalk, and consists of a head of yellow florets surrounded by 2 or 3 rows of green bracts which comprise the involucre. The florets are of 2 kinds, those forming the central disc being bisexual and tubular, while those in the single marginal row are female and ligulate. The plant flowers from April to May or early June.

Pollination. As in *Taraxacum*, the stamens of *Doronicum* mature before the styles, but the pollination mechanism is different. The anthers in *Doronicum* also form a cylinder, but when they dehisce the pollen falls on to the flattened tip of the style formed by the ends of the stigmas that are pressed closely together. As the style lengthens it acts as a piston, pushing the pollen upwards and making it accessible to pollinating insects. After the style has emerged from the cylinder the stigmas separate, as in *Taraxacum*, causing the previously hidden surfaces to be exposed and ready to receive pollen from other flowers.

Alternative flowers for study. Alternatives to *Doronicum* are numerous. Native British plants include *Tussilago farfara* (Colt's-foot), which flowers in March and April, and *Senecio jacobaea* (Ragwort), which is in flower from June until October. Garden plants such as the single forms of *Aster*, *Dahlia* and *Dendranthema* could also be used.

A. A flower-bud, showing the involucre of bracts in 2 rows protecting the young florets.
Bud: 8 x 8 mm

B. L.S. of the capitulum, consisting of regular, bisexual disc-florets in the centre and irregular, female ray-florets at the margin. Note the convex receptacle.
Receptacle: 10 mm in diameter
Inflorescence: 6.2-6.4 cm in diameter
Bracts: 10-12 x 2 mm

C. A single disc-floret, showing the inferior ovary and the 5-lobed corolla-tube with protruding stigmas. The corolla of each disc-floret is surrounded by a pappus of long hairs, considered to be a modified form of calyx.
Corolla: 5.25 mm Pappus: 3.5 mm

D1. A ray-floret. The corolla surrounds a style that usually branches into 2 stigmas but may be unbranched (see D2). The pappus is absent from ray-florets. Note the hairs on the corolla-tube.

D2. Upper portion of a style with an unbranched stigma.

E. L.S. of a disc-floret. The filaments are short and are borne separately on the corolla-tube. The anthers are 2-celled, face inwards, and are united to form a tube.
Filament: 3 mm Anther: 2.65 mm
Style: 5 mm Stigma: 0.4 mm

F. T.S. of ovary showing the single ovule and the fluted external wall of the ovary formed from 2 united carpels.

G. L.S. of ovary with its single, basal ovule.
Ovary: 2 x 0.7 mm Ovule: 1 mm

H. A fruit, crowned by the pappus, which has developed from one of the disc-florets. Fruits of this hybrid are usually sterile.
Cypsela: 1.5 mm

19.3 Centaurea scabiosa L. Greater Knapweed

Distribution. Native of roadsides, hedgebanks, cliffs and dry grassland throughout lowland Britain, continental Europe, and W. Asia.

Vegetative characteristics. An erect, perennial herb with a stout, branching, woody stock, and grooved stems bearing deeply lobed leaves.

Floral formula.
 Fertile florets: K(pappus) C5 A(5) $\overline{G}(2)$

Flower and inflorescence. The stem reaches a height of 30-90 cm, widening at the top below the inflorescence. The numerous red-purple florets appear from July until September.

Pollination. The sterile, larger, outer florets that surround the fertile disc-florets render the inflorescence more conspicuous. Bees and various flies seek out the nectar which is secreted at the base of the narrow corolla-tube. The nectar rises up the corolla-tube, and becomes more accessible to insect-visitors. The genus

Centaurea provides notable examples of secondary pollen presentation. The anthers of the fertile disc-florets form a tube and dehisce inwards before the flower opens. The style grows up through this tube, and the stiff hairs situated below the stigma sweep the pollen to the top of the tube where the connectives joining the anthers end in a ring of triangular teeth. The flower then opens, and the white pollen-masses protrude from the tube. When a visiting insect touches the sensitive anthers, it causes them to contract and deposit pollen on to its body. The female stage occurs after the pollen is dispersed, when the style grows until it is well beyond the anther-tube. The 2-branched tip of the style bends outwards, exposing the inner surface of the receptive stigma to pollen-laden insects.

Alternative flowers for study. C. nigra (Lesser Knapweed) would be a suitable alternative, differing from *C. scabiosa* in having a smaller capitulum and usually entire leaves. Comparison should be made with *C. cyanus* (Cornflower).

A. A flowering stem that widens out to support the capitulum. One of the capitula is in bud while the other is in flower and shows numerous fertile disc-florets and an outer whorl of larger, sterile ray-florets. On the lower part of the capitulum are series of overlapping bracts.

B. L.S. of bud, showing bracts and florets attached to the flattened receptacle.
Bud: 15 x 13 mm

C. The outer surface of a bract. The deeply toothed portion is blackish brown and the lower part is green. The bracts elongate during the fruiting stage.
Bract (shape and size variable): 7-8 x 4-5 mm

D. A sterile ray-floret. Note the rudimentary pappus attached to the infertile fruit.
Corolla-tube: up to 20 mm
Corolla-lobes: up to 10 mm

E. A fertile floret. The corolla has 5 free lobes arising from a united swollen portion that tapers at its base into a narrow tube.
E1: Whole floret in bud. E2: Upper portion of floret at male stage, with pollen-grains adhering to the ring of hairs on the style.
E3: Late female stage. The style has lengthened and all the pollen has been removed. Note the fruit and bristly pappus in E1 and E3.
Floret in bud: 18 mm

Floret at male stage
Corolla-lobes: 9 mm Style protruding 4 mm
Floret at female stage
Corolla-tube to swelling: 13 mm
Corolla-lobes: 10 mm Style protruding 6 mm

F1. Corolla cut away to reveal the anther-tube.
F2. Corolla-tube and anther partly cut away to expose the style.
F3. Detail of 2 of the anthers at dehiscence. Most of the pollen has been removed for clarity. A pointed connective joins the anthers together.
Anther: 6 x 1-1.5 mm Filament: 4 mm

G. The style, projecting from the anther-tube.
G1: Late male stage, showing pollen grains on the hairs of the style. G2: Female stage. The pollen has gone, and rings of long and short hairs can be seen below the stigma-branches, now wide open to receive pollen from another flower.
Style at late male stage protruding up to 3 mm
Style at female stage protruding up to 5 mm

H. L.S. of fruit, a cypsela. The single seed can be seen within the carpel wall.

I. The cypsela has a pappus of bristly hairs, each one covered with upward-pointing teeth.
Fruit: 4 x 1.5 mm Pappus: 5 mm

A

B

C

D

E1 E2 E3

F1 F2 F3 G1 G2

H

I

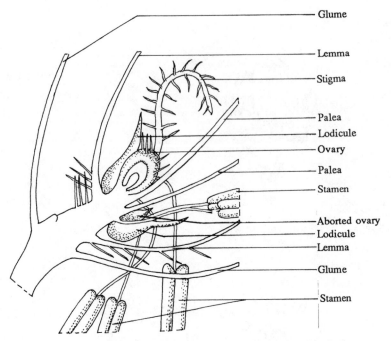

Glume

Lemma

Stigma

Palea
Lodicule
Ovary
Palea

Stamen

Aborted ovary
Lodicule
Lemma

Glume

Stamen

L.S. of flowering spikelet of *Arrhenatherum elatius*

Palea

Stigma

Rachilla

Ovary

Filament

Anther

Glume

Lemma

Lodicule

Structure of bisexual floret of a typical grass

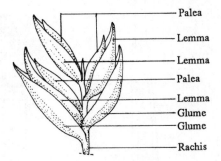

Palea

Lemma

Lemma

Palea

Lemma
Glume
Glume

Rachis

Structure of a spikelet containing
3 florets with 2 glumes at the base

Spikelet scar

Bristle-like hairs

Portion of rachis of *Triticum*, Wheat, with spikelets
removed to show zig-zag arrangement

Fig. 61. Flower structure of grasses.

20 GRAMINEAE Juss. (Poaceae Barnhart)

737 genera and 7950 species

Grass family

Distribution. World-wide.

General characteristics. Annual or perennial herbs with fibrous roots, forming tufts or mats. Stems (culms) simple or branched, erect or lying on the ground, usually cylindrical, jointed, usually hollow except at the joints (nodes). Leaves alternate in 2 rows on opposite sides of the stem, originating at the nodes, often crowded at the base, and consisting of sheath, ligule and blade; sheath with margins free and overlapping, or more or less united to form a tube round the stem; ligule (at junction of sheath and blade) membranous, sometimes reduced to a fringe of hairs or rarely absent; blade (lamina) usually long, narrow, and passing imperceptibly into the sheath. Veins normally parallel. Inflorescence consisting of numerous spikelets which form spikes, racemes, or panicles. Flowers usually bisexual, small and inconspicuous. Stamens usually 3; anthers 2-celled. Ovary superior, of 3 united carpels, unilocular, containing 1 ovule; styles usually 2; stigmas generally feathery. Fruit 1-seeded, usually a caryopsis, rarely a nut or berry. Endosperm abundant.

Economic and ornamental plants. Economically the Gramineae are probably of greater importance than any other family. Plants providing food for man include *Triticum* spp.(Wheat), *Avena sativa* (Oats), *Hordeum vulgare* (Barley), *Secale cereale* (Rye), *Zea mays* (Maize, termed 'Corn' in the United States), *Oryza sativa* (Rice), *Panicum miliaceum* (Common Millet), *Sorghum vulgare* (Sorghum), and *Saccharum officinarum* (Sugar Cane). Many others are grown as fodder for domestic animals. *Phragmites australis* (*P. communis*) (Common Reed) is used for thatching, and *Ammophila arenaria* (Marram Grass) is valuable for binding and consolidating drifting sand. The various genera collectively known as Bamboo have numerous economic uses, especially in Asian countries. Many grasses are used in lawns and other turfed areas, and others are grown for ornament in the garden, or dried and used for floral decoration.

Classification. The Gramineae form one of the largest families of flowering plants. Recent studies show that the family may be divided into 7 subfamilies and over 50 tribes on a wide variety of characters. Some of the principal genera are listed here in their appropriate subfamily.

I. Bambusoideae
 Arundinaria (30) Madagascar, S. and E. Asia, America.
 Oryza (19) tropics.

II. Centothecoideae (a few small genera native in tropical rain forests)

III. Arundinoideae
 Cortaderia (24) S. America, New Zealand.
 Phragmites (4) world-wide.
 Arundo (3) Mediterranean region, Taiwan.

IV. Chloridoideae
 Eragrostis (250) world-wide.
 Spartina (17) mainly temperate regions.

V. Panicoideae
 Panicum (600) ⎫
 Paspalum (250) ⎪
 Digitaria (170) ⎪
 Setaria (125) ⎬ tropical and warm temperate regions.
 Pennisetum (70) ⎪
 Saccharum (30) ⎪
 Sorghum (24) ⎭
 Zea (4) Central America.

VI. Pooideae
 Festuca (300) world-wide.
 Poa (250) temperate and cold regions.
 Agrostis (120) world-wide.
 Bromus (100) mainly temperate regions.
 Elymus (100) N. temperate region.
 Melica (70) temperate regions.
 Agropyron (40) temperate regions.
 Alopecurus (25) N. temperate region.
 Triticum (20) Europe, W. Asia.
 Hordeum (20) N. temperate region.
 Avena (15) Old World temperate regions.
 Phleum (15) temperate Eurasia, America.
 Briza (12) temperate regions.
 Holcus (9) Eurasia, N. and S. Africa.
 Lolium (8) temperate Eurasia.
 Arrhenatherum (6) Europe, N. Africa, W. Asia
 Secale (3) Eurasia.
 Dactylis (1) temperate Eurasia.

VII. Stipoideae
 Stipa (150) tropical and temperate regions.

20.1 Arrhenatherum elatius (L.) J. & C. Presl False Oat-grass

Distribution. Native in N. Africa, W. Asia, and most of Europe including the British Isles, and introduced into N. America, Australia and New Zealand. It is a common constituent of rough grassland and is also found in hedgerows and on waste ground.

Vegetative characteristics. A tufted, usually glabrous perennial, with more or less erect stems 50-150 cm in height. The leaves are finely pointed, flat and rough, and the ligules short and membranous.

Floral formula. P2 A3 G̲(3)

Flower and inflorescence. The inflorescence is a panicle with about 10 nodes from which spring clusters of 4-6 stalks bearing the pedicels of the spikelets. Each spikelet normally consists of 2 florets (occasionally 3 or 4), the upper one bisexual and the lower male. The flowering period extends from June until September or even as late as November if the weather is suitable. The flowers usually open between 5 and 7 a.m.

Pollination. Like other grasses, *A. elatius* is well adapted to wind-pollination, having pendulous stamens easily moved by currents of air, abundant pollen, and feathery stigmas presenting a large surface to the wind-borne pollen from other plants.

Alternative flowers for study. Any species of the closely related genus *Avena*, including *A. sativa* ssp. *fatua* (Wild Oat), would be a suitable alternative. Floral variations within the Gramineae are many, and include shape of inflorescence, arrangement of spikelets, and number of florets in a spikelet.

A. A single spikelet. At its base are the 2 glumes (see C1 and C2). Above these are the 2 lemmas (see D1).

B. A spikelet at flowering time. The upper (bisexual) floret is fully open and its 3 stamens are hanging down to shed their pollen. One of the 2 feathery stigmas protrudes just below the short-awned lemma that is backed by the upper glume. Two paleas can be seen in the centre, one from each floret. The rest of the lower (male) floret is hidden by the long-awned lemma that is backed by the lower glume.

C1. The single-nerved, lower (outer) glume.
Lower glume: 4-5.5 x 1 mm

C2. The 3-nerved, upper (inner) glume.
Upper glume: 7.5-8 x 3 mm

D. A male floret. D1: The 7-nerved lemma with its long awn. In contrast, the lemma of the bisexual floret is awnless or only slightly awned and may have dorsal bristles (see B). D2: The hairy-edged palea, which is angular and readily fits into the lemma forming a protective chamber.
Long-awned lemma: 8 x 2 mm Awn: 14-17 mm
Palea: 7 x 1 mm

E. L.S. of a spikelet, showing the 2 florets borne on a short rachilla. The superior ovary of the upper (bisexual) floret contains a single ovule and is crowned by 2 styles with feathery stigmas (one only shown). At its base is one of the 2 lodicules which are considered to be vestiges of a perianth. The 3 stamens are inserted on the rachilla between the ovary and the lodicules. The lower (male) floret exhibits an abortive ovary. Two of the 3 stamens are shown. The long, narrow structures in both florets are sections of the lower portions of the glumes, lemmas and paleas. At maturity, the part of the rachilla above the glumes breaks away, carrying with it both male and bisexual florets.

F. Detail of the 2-celled anther which dehisces lengthwise, liberating a little of the pollen at a time from the boat-shaped cavity.
Filament: 5 mm Anther: 4 mm

G. A bisexual floret with stamens removed to reveal the superior ovary ending in 2 styles, and the 2 lodicules which become swollen with sap and cause the flower to open. Below the lodicules is a portion of the lemma and behind the ovary is the lower part of the palea.
Ovary: 2 x 0.8 mm Styles: 3-4 mm
Lodicules: 1.5-2.25 mm

H. Part of the stigma, showing some of the minute papillae which have caught 2 pollen grains.

I. The fruit, a caryopsis, often called the 'seed' or 'grain'. The hairy outer surface is, in fact, the pericarp, which adheres to the testa (seed-coat). The true seed becomes visible only when the pericarp has been removed.
Fruit: 4.75 x 1.75 mm

20.1 *Arrhenatherum elatius*, False Oat-grass

20.2 *Zea mays*, Maize, Sweet Corn

20.2 **Zea mays** L. Maize, Sweet Corn, Indian Corn, Corn

Distribution. Zea mays has been grown as a crop plant in warm regions of America for thousands of years, but the plant has never been found in the wild. It was introduced into Europe by Columbus, and is now cultivated also in Asia and Africa.

Vegetative characteristics. A stout annual grass with a solid stem 1-4 m high or more, bearing long, linear-lanceolate, sheathing leaves. Adventitious roots develop from the lowest nodes of the stem forming whorls of 'brace' or 'prop' roots.

Floral formula. Male: P2 A3 Female: P0 G(3)

Flower and inflorescence. Male and female inflorescences occur separately on the same plant. The male inflorescence consists of a panicle or 'tassel', while the female inflorescence is a modified spike, the 'ear'. This is enveloped in protective bracts called 'husks', from the top of which protrude long, thread-like styles or 'silks'. After fertilisation the

inflorescence develops into a closely packed collection of fruits, the 'kernels', which form the 'cob'. Flowering commences in June or July and the cobs begin to ripen in September.

Pollination. The male flowers ripen first, and the anthers start shedding large quantities of pollen at sunrise. The pollen is viable for about 24 hours and is carried by the wind and by gravity on to the long styles ('silks') of the female inflorescence. After pollination the silks dry up, while those that have not been pollinated grow very long and remain swollen. Fertilisation takes 12-28 hours and during this time the pollen-tube may have travelled a distance of 25 cm.

Alternative flowers for study. Zea mays is a very distinctive grass and readily available. There is considerable variation within the species, shown by the numerous cultivars, some with striped leaves or grains of various colours.

A. A branch of the male inflorescence. Some of the pairs of spikelets have been removed for clarity.

B. A young male spikelet, showing the 2 hairy glumes that arise from a slightly twisted rachilla. Two stamens can be seen emerging from between the glumes.
Male spikelet: 8 x 3 mm

C. A male spikelet with the 2 glumes drawn apart to reveal the 2 florets. Each floret has an outer, 3-nerved, concave lemma and an inner, 2-nerved, almost flat palea with incurved margins, and 3 stamens each with a short filament that elongates after flowering. The upper floret within each pair of glumes matures first.
Lower glume: 7 x 3 mm Upper glume: 6 x 3 mm
Lemma: 5.9 x 3 mm Palea: 5-6.5 x 2 mm

D. A bilobed anther with part of its filament. The anthers may be green, yellow, pink, or purple, and shed copious pollen through a slit at the tip.
Filament: up to 6 mm Anther: 5 x 1 mm

E. L.S. of female inflorescence, at the base of which are closely spaced nodes bearing modified leaves, the 'husks'. Numerous female florets surround the well-developed axis, their long styles projecting above the husks.
Cob: 12.5 x 2.5 cm Styles: up to 3.7 cm
Leaf-sheaths + blades: up to 43 x 5 cm

F. L.S. of a female spikelet. The ovary ends in a long style, only a portion of which is shown. The female spikelet is basically 2-flowered, but 1 floret usually aborts. Two glumes and 2 lemmas can be seen at the base of the ovary. Ovary (at early stage): 2 x 2 mm

G. L.S. of a fruit removed from the cob. The fruit, a caryopsis, is 1-seeded and indehiscent with a persistent pericarp. The seed is usually without a testa, but has a well-developed starchy endosperm (stippled in drawing) and a large embryo (white in drawing). The base of the deciduous style is just visible on the pericarp-wall.

H. The whole fruit. The lower, elongated portion shows where it was attached. The pericarp, or 'hull', is usually smooth but in some varieties it may be wrinkled. The fruits are arranged in even rows down the cob, the number of rows varying according to the cultivar concerned.
Fruit: 10 x 10 mm

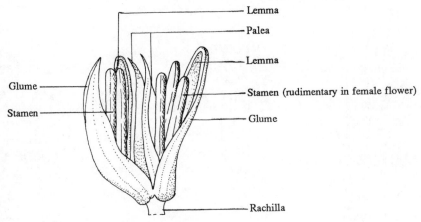

Spikelet of male flower before dehiscence

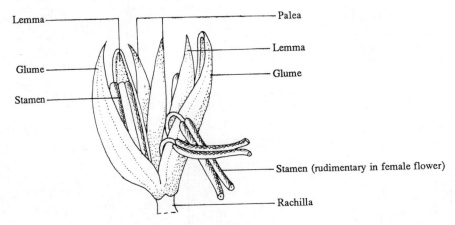

Spikelet of male flower after dehiscence

Female flower, showing all the parts that may be present

Fig. 62. Flower structure of *Zea mays*.

| Gramineae
Triticum hybrid, Bread Wheat | *Juncaceae *Juncus*,
Rush and *Luzula*, Wood-rush | *Cyperaceae *Carex acutiformis*,
Lesser Pond-sedge |

Gramineae *Triticum* hybrid, Bread Wheat

Flower (floret)

Flower-spike

Node

Internode

Leaf-sheath

Lower portion of shoot

Hollow internode of stem

Ligule

Open sheath

Leaf-blade

Stem section and leaf-sheath

***Juncaceae *Juncus*, Rush and *Luzula*, Wood-rush**

Flower of *Juncus*

Inflorescence of *Juncus*

Flower of *Luzula*

Basal tuft of leaves

Flowering plant of *Luzula*

Stem with pith

Stem section of *Juncus*

***Cyperaceae *Carex acutiformis*, Lesser Pond-sedge**

Male inflorescence

Female inflorescence

Flowering shoot

Male flower

3-angled stem

Female flower

Leaf-blade

Ligule

Closed sheath

Stem section and leaf-sheath

Fig. 63. Distinguishing characters of grasses, rushes and sedges.

Spathe

Upper portion of spadix

Sterile male flowers

Fertile male flowers
Sterile female flowers
Fertile female flowers

Fig. 64. L.S. of inflorescence of *Arum maculatum*.

21.1 *Arum maculatum*, Lords-and-ladies, Cuckoo-pint

21 ARACEAE Juss. Arum family

106 genera and 2950 species

Distribution. Mostly tropical and subtropical, but a few in temperate regions.

General characteristics. Mainly herbs with tubers or rhizomes, occasionally shrubs, sometimes climbing or epiphytic with aerial roots, more rarely marsh or water plants. Poisonous latex often present. Leaves parallel-veined or net-veined, alternate. Flowers without bracts, usually massed together to form a more or less cylindrical spadix enclosed in a large spathe; bisexual or unisexual. Perianth present or absent. Stamens usually fewer than 6, often united; staminodes often present. Ovary of varied structure, frequently reduced to one carpel, superior or embedded in spadix. Fruit a berry. Endosperm present or absent. Flowers usually with a disagreeable smell.

Economic and ornamental plants. The chief economic plant is *Colocasia esculenta* (Taro), whose large, starchy tubers are an important tropical root crop. Species of *Alocasia* and *Xanthosoma* are also cultivated for their roots or leaves in tropical regions, and *Monstera deliciosa* for its edible fruit. In colder climates, *Monstera* is grown as a house plant on account of its curious leaves. Other house plants include *Caladium*, *Aglaonema*, *Philodendron*, *Syngonium*, and *Dieffenbachia*, grown for their leaves, and *Anthurium* and *Spathiphyllum* for their flowers. *Zantedeschia* (Calla Lily) and *Dracunculus* (Dragon Arum) can be grown outside in favourable areas, and some species of *Arisaema* and *Lysichiton* are quite hardy.

Classification. The grouping of the Araceae is very difficult, and account has to be taken of a variety of characters. The following classification into 8 subfamilies is based on that of Engler (see References):

I. Pothoideae (land plants; no latex or raphides; flowers usually bisexual)
 Anthurium (700) tropical America, W. Indies

II. Monsteroideae (land plants; no latex; raphides present; flowers bisexual)
 Spathiphyllum (36) tropical America.
 Monstera (22) tropical America, W. Indies.

III. Calloideae (land or marsh plants; latex present; leaves never arrow-shaped, usually net-veined; flowers usually bisexual)
 Lysichiton (2) N.E. Asia, western N. America.
 Calla (1) N. temperate and subarctic region
 Symplocarpus (1) temperate E. Asia, eastern N. America.

IV. Lasioideae (land or marsh plants; latex present; leaves arrow-shaped, net-veined; flowers bisexual or unisexual)
 Dracontium (13) tropical America.

V. Philodendroideae (land or marsh plants; latex present; leaves usually parallel-veined; flowers unisexual)
 Philodendron (500) tropical America.
 Zantedeschia (6) tropical and S. Africa.

VI. Colocasioideae (land or marsh plants; latex present; leaves net-veined; flowers unisexual, stamens united)
 Alocasia (70) tropical Asia.
 Xanthosoma (45) tropical America.
 Colocasia (6) tropical Asia.

VII. Aroideae (land or marsh plants; latex present; leaves net-veined; flowers usually unisexual, stamens free or united)
 Arisaema (150) E. Africa, tropical Asia, eastern N. America.
 Arum (15) Europe, Mediterranean region.
 Spathicarpa (6) tropical S. America.
 Dracunculus (3) Mediterranean region.
 Sauromatum (2) Old World tropics.

VIII. Pistioideae (floating plants; no latex; flowers unisexual)
 Pistia (1) tropics and subtropics.

21.1 Arum maculatum L. Lords-and-Ladies, Cuckoo-pint

Distribution. Generally distributed throughout the British Isles, though less common in Scotland. It is also native in Europe northwards to southern Sweden and in N. Africa. It prefers shady places such as hedgerows and woods, and can become a persistent weed in garden shrubberies. In the United States it can be grown as far north as New York City.

Vegetative characteristics. An erect, glabrous, perennial herb with a tuberous rootstock and long-stalked, spear-shaped leaves which are often blackish-spotted. All parts of the plant contain an acrid juice which may irritate the skin.

Floral formula: Male: P0 A2-4 Female: P0 G1

Flower and inflorescence. The inflorescence, which appears in April and May after the leaves, is a club-shaped structure known as the spadix. The upper portion is dull purple in colour; the lower portion bears whorls of small, unisexual flowers, the male above, the female below. The spadix is enclosed by the spathe, a large, yellowish green bract sometimes spotted with purple.

Pollination. The highly specialised structure described above forms a pitfall trap. The female flowers mature first, and when they are in a receptive state the spadix emits a smell (and incidentally generates heat) which attracts small flies. If the insects alight on the spadix or the inner surface of the spathe, both of which are smooth and slippery, they lose their grip and fall through the ring of bristles attached to the sterile male flowers to the bottom of the pit. Here they come into contact with the mature female flowers and transfer to them any pollen they may be carrying from another plant. After pollination, the stigmas soon wither and the fertile male flowers situated above ripen and shower the trapped insects with pollen. The bristles of the sterile male flowers then wither and the surface of the spadix itself becomes wrinkled enabling the flies to climb up out of their prison and fly off to act as agents in the cross-pollination of another plant.

Alternative flowers for study. A number of Araceae are grown indoors for their flowers, but it should be noted that there is often considerable variation between one genus and another.

A. The complete flowering shoot on a short stem. The enveloping lower portion of the spathe conceals the reproductive organs, while the upper part of the spadix is clearly visible against the open portion of the spathe.
Spathe: 17-25 x 6.5-10.5 cm

B. Part of the spathe has been cut away to reveal the pitfall trap formed by the lower portions of the spathe and spadix. Just below the constricted part of the spathe are the sterile male flowers with long bristles which exclude the larger insects; then come the fertile male flowers, below these a small number of sterile female flowers, also provided with bristles, and at the base the fertile female flowers. Both male and female flowers are devoid of perianth-segments.
Lower (flower-bearing) part of spadix: 21-26 mm
Upper part of spadix: 43-60 mm

C. A single, fertile male flower, prior to dehiscence, with portion of spadix attached.
Male flower: *c.* 1 x 1 mm
Bristle of sterile male flower: 4-6 mm

D. A male flower, showing pollen emerging from a slit in the stamens.

E. L.S. of a portion of the male inflorescence, showing the paired nature of the stamens and the pollen grains inside them.

F. A fertile, female flower with sessile stigma.
Female flower (at flowering time): 3 x 2.5 mm
Stigma: *c.* 0.9 mm in diameter

G. T.S. of unilocular ovary showing the ovules attached to the placenta.

H. L.S. of ovary with portion of spadix attached. One ovule is still connected to the placenta, the other has been slightly displaced in sectioning.
Loculus of ovary: *c.* 0.5 mm

I. A sterile female flower with its long, hooked bristle.
Swollen portion: 2 mm Bristle: 3-5 mm

J. The infructescence, which has developed from the individual carpels. The scarlet, few-seeded fruits are partly enveloped by the remains of the withered spathe.
Infructescence: 30-40 x 18-19 mm
Single berry: 8 x 8 mm to 9 x 7 mm

22.1 *Lilium martagon*, Martagon Lily, Turk's-cap Lily

22 LILIACEAE Juss. Lily family

220 genera and 3500 species

Distribution. World-wide.

General characteristics. Mostly perennial herbs with sympodial rhizomes or bulbs, some succulents (e.g., *Aloe*, *Gasteria*) and climbers (e.g. *Smilax*, *Gloriosa*). Leaves basal or cauline, alternate or whorled, sometimes fleshy or spiny-margined. *Ruscus* exhibits cladodes. Inflorescence most commonly racemose, but cymose in the apparent umbels of *Allium*, *Agapanthus*, etc. Solitary terminal flowers occur in *Tulipa* etc. Flowers usually bisexual, regular. Perianth-segments in 2 whorls of 3, free or united, petaloid or sometimes sepaloid. Stamens usually in 2 whorls of 3, with anthers usually facing inwards. Ovary superior, of 3 united carpels, usually 3-locular, with 2 rows of numerous, anatropous ovules on axile placentas in each loculus, rarely unilocular with parietal placentation. Fruit usually a capsule, sometimes a berry. Seed with abundant endosperm. Flowers usually insect-pollinated.

Economic and ornamental plants. The family contains a few plants of economic importance. *Asparagus officinalis* (Asparagus), and various species of *Allium*, known to the vegetable gardener as Onions, Leeks, Shallots, Chives, Garlic, etc. are cultivated as food or for flavouring, while *Aloe*, *Colchicum*, *Urginea* (Squill), and *Veratrum* are of medicinal value. By contrast, many plants are highly decorative and are often grown in gardens. Most of the genera listed in the next section are in this category.

Classification. The Liliaceae is considered here in a broad sense, but some authorities restrict the family to *Lilium* and closely related genera, placing other genera in a number of separate families. An indication of this treatment is shown by the grouping below.

1. Liliaceae in a strict sense:

 Lilium (100) N. temperate region.
 Tulipa (100) temperate Eurasia.
 Fritillaria (99) N. temperate region.
 Gagea (70) temperate Eurasia.
 Erythronium (21) N. America, Eurasia.

2. The Hyacinth group:

 Ornithogalum (120) Europe, W. Asia, S. Africa.
 Urginea (100) Mediterranean region, Africa.
 Lachenalia (80) S. Africa.
 Muscari (60) Europe, W. Asia, N. Africa.

 Scilla (40) temperate Eurasia, S. Africa.
 Hyacinthella (16) S.E. Europe, S.W. Asia.
 Camassia (6) N. and S. America.
 Chionodoxa (5) E. Mediterranean region.
 Hyacinthoides (4) W. Europe, N. Africa.
 Galtonia (3) S. Africa.
 Hyacinthus (3) W. and C. Asia.
 Puschkinia (2) Turkey, Caucasus.
 Veltheimia (2) S. Africa.

3. The Aloe group:

 Aloe (363) tropical and S. Africa, Arabia, Madagascar.
 Haworthia (68) S. Africa.
 Kniphofia (65) E. and S. Africa, Madagascar, Arabia.
 Gasteria (13) S. Africa.
 Asphodelus (12) Mediterranean region to Himalaya.

4. The Onion group:

 Allium (700) N. hemisphere.
 Ipheion (25) S. America.
 Brodiaea (15) western N. America.
 Agapanthus (9) S. Africa.

5. The Lily-of-the-valley group:

 Polygonatum (55) N. temperate region.
 Ophiopogon (50) Himalaya to E. and S.E. Asia.
 Convallaria (3) N. temperate region.

6. Well-known genera not included in the above groups:

 Smilax (200) tropical and temperate regions.
 Asparagus (100) Old World.
 Colchicum (65) Eurasia, N. Africa.
 Calochortus (60) western N. America.
 Hosta (40) China, Japan, Korea.
 Trillium (30) W. Himalaya to Japan, N. America.
 Veratrum (25) N. temperate region.
 Hemerocallis (15) C. Europe to Japan.
 Tricyrtis (11) Himalaya to Japan.
 Ruscus (6) Azores to Iran.
 Lapageria (1) Chile.

22.1 Lilium martagon L. Martagon Lily, Turk's-cap Lily

Distribution. Lilium martagon is the most widespread species of lily, and is found growing wild from southern Europe to Japan, usually on the lower wooded slopes of mountains or in grassy glades and meadows. It has been introduced into many other countries, and is a common garden plant, often escaping and becoming naturalised.

Vegetative characteristics. An erect, hardy perennial with an ovoid, yellow bulb made up of numerous overlapping scales. The stout stem, 50-100 cm tall, bears whorls of 5-10 obovate-lanceolate leaves.

Floral formula. P3+3 A3+3 G(3)

Flower and inflorescence. The plant produces a large, terminal raceme of bisexual, regular, pendulous flowers during August and September. The flowers are typically dull purple or purplish red with darker, raised spots, but other colour forms, ranging from dark purple to white, are in existence. Their scent is often strong and unpleasant.

Pollination. The flowers are pollinated mostly by night-flying insects, usually hawk-moths. The moths are attracted by the scent of the nectar, which becomes much stronger in the evening. The nectar is produced in a groove at the base of each of the perianth-segments. The edges of this groove are flanged, and fringed with reddish hairs, forming a tube. A visiting moth may land on the perianth or may hover around the flower, inserting its proboscis into one nectar-groove after another. While the moth is foraging for nectar, it dusts the stigma with pollen from another flower, and also receives fresh pollen. Self-pollination may take place when the last remaining pollen-grains fall on to the style which has bent round, bringing the stigma closer to the anthers.

Alternative flowers for study. The flowers of other species of *Lilium* may vary considerably in appearance, though their structure remains uniform. In the genus *Ruscus*, the flowers are unisexual and are borne from January to April on the midrib of a leaf-like cladode; the male flowers have only 3 stamens, and the fruit is a berry. Other genera, e.g. *Muscari* and *Kniphofia*, have the perianth-segments united, and in some species of *Allium*, e.g. the native *A. oleraceum* and the naturalised *A. paradoxum*, bulbils are present amongst the flowers.

A. A young flower. The pedicel of the flower has 2 leaf-like bracts at its base. Each of the 6 perianth-segments has, in its lower part, a central nectar-secreting groove flanged on each side.
Pedicel: 38 mm Bracts: 18-23 x 3-8 mm
Perianth-segments: 32 x 12 mm
Nectar-groove: 15 x 1 mm

B. Two perianth-segments have been removed entirely, and 2 more cut off near the base to show the stamens attached to the receptacle. Each stamen stands opposite a perianth-segment. The superior ovary extends into a long, curved style.
Stamen: 24 mm Ovary: 8 mm
Style + stigma: 25 mm

C. A young anther, dehiscing lengthwise to shed the abundant pollen. The 2-celled, versatile anthers are yellowish orange or reddish purple, and are supported by long, pinkish filaments.
Anther: 7-8 mm

D. L.S. of ovary and lower part of style, showing numerous ovules attached to the axile placenta. At the base of the ovary are the lower parts of 2 stamens and 2 perianth-segments.

E. T.S. of ovary. In each of the 3 loculi, the ovules form a double row extending the whole length of the ovary.
Ovary: 1.75 mm in diameter

F. The more or less 3-lobed stigma, showing the amount of lobing visible from 2 sides, and the numerous small papillae.
Stigma: 2 mm across

G. Young fruit, still green and fleshy. The fruit develops into a capsule, which splits into 3 parts to distribute the large, flat seeds. After dehiscence the capsule becomes dry and papery in texture and remains for a long time on the dead stem.
Young fruit: 32 x 10 mm
Mature fruits: 10-34 x 8-18 mm

22.2 Hyacinthoides non-scripta (L.) Rothm. Bluebell
(Endymion non-scriptus (L.) Garcke)

Distribution. Common in woods and hedgerows throughout the British Isles. Native also in Belgium, the Netherlands, and France. Doubtfully native in the Iberian peninsula, N. Italy, and N.W. Germany. It is frequently cultivated in European and American gardens.

Vegetative characteristics. A glabrous herb with several linear leaves and a flowering scape arising from an ovoid bulb.

Floral formula. P3+3 A3+3 G(3)

Flower and inflorescence. The violet-blue (rarely pink or white) flowers form a one-sided, racemose inflorescence from April to June. They are erect in bud but nodding when fully open. The tips of the perianth-segments are somewhat recurved, and the anthers are cream in colour.

Pollination. Bumble-bees, hover-flies and other insects seek out the nectar secreted by glands in the ovary wall, an arrangement common to many members of the Liliaceae. The removal of nectar would normally leave the flower undamaged, but insects have been seen to bite through the base of the perianth-tube in order to reach the nectar. This 'illegitimate' behaviour, usually perpetrated by powerful but short-tongued bumble-bees occurs quite commonly in connection with tubular flowers.

Alternative flowers for study. H. hispanica (Spanish Bluebell), which flowers in May, differs from H. non-scripta in having blue anthers, and paler, more spreading perianth-segments which are not recurved.

A. A flowering plant. An ovoid whitish bulb, which is renewed annually gives rise to several glabrous, linear leaves and a tall flowering scape. The scape bears a one-sided raceme with unopened buds erect and clustered near the apex while the fully open lower flowers are nodding. Inflorescence: 13 cm

B. An individual flower with 2 bracts at the base of the curved pedicel. The 6 perianth-segments are in 2 whorls and are united at their base. Bracts: 10-12 mm Pedicels: 5-10 mm

C. L.S. of the regular, bisexual flower exposing the superior ovary and 4 of the 6 stamens. The anthers are versatile and are attached at their back to the filaments. Perianth-segments: 13-17 x 4.5 mm

D1. A young inner stamen from a well-developed flower-bud. The filament is the same colour as the perianth-segment and is joined to it. The cream-coloured anther is 2-celled, faces inwards and dehisces lengthwise. The anthers are attached to the perianth in 2 rows, the outer row to the middle of the segments and the inner row nearer the base. Filament (outer): 6 mm (inner): 3 mm Anther: 3.75-4 mm

D2. Side view of stamen showing how the anther is attached by its back to the filament.

E. T.S. of the 3-locular ovary composed of 3 united carpels. The ovules are arranged in 2 rows in each loculus. Placentation is axile. Ovary: 4 mm in diameter Ovules: 0.35-0.65 mm

F. L.S. of ovary exposing the rows of ovules in 2 of the 3 loculi. The ovary ends in a long, simple style. Ovary: 3.25-4 mm Style + stigma: 4-5 mm

G. Detail of the receptive papillose stigma.

H. The green, fleshy fruit before dehiscence, a 3-lobed capsule with the persistent perianth at its base.

I. The fruit at dehiscence, now papery and straw-coloured, opening by 3 valves to disperse the large black seeds. Fruit: 10-16 x 8-14 mm

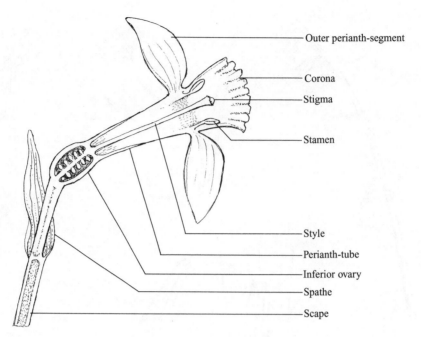

Outer perianth-segment

Corona

Stigma

Stamen

Style

Perianth-tube

Inferior ovary

Spathe

Scape

Fig. 65. L.S. of *Narcissus* flower.

Hybrid *Narcissus*, Daffodil

N. bulbocodium,
Hoop-petticoat Daffodil

N. cyclamineus

N. jonquilla,
Jonquil

N. poeticus,
Poet's Narcissus

N. triandrus,
Angel's Tears

Fig. 66. Flowers of various *Narcissus*.

23 AMARYLLIDACEAE Jaume St-Hil. Daffodil family

70 genera and 850 species

Distribution. Mostly tropical or warm regions.

General characteristics. Usually found in dry areas, often bulbous, leafing only in spring or in the rainy season; many with rhizomes. Leaves more or less linear, mostly basal. Inflorescence usually on a scape, with one or more spathes, cymose, but often condensed and resembling an umbel or head; sometimes flowers solitary. Flowers bisexual, regular or irregular. Perianth-segments in 2 whorls of 3, free or united, petaloid. Stamens in 2 whorls of 3, sometimes represented by staminodes; anthers 2-celled, usually facing inwards, versatile or occasionally attached at the base. Ovary inferior, of 3 united carpels, 3-locular or rarely unilocular; style single, with 3-lobed or capitate stigma; placentation axile; ovules numerous, anatropous. In some genera, e.g. *Narcissus*, a conspicuous corona, looking like an extra whorl of perianth-segments, is situated between the normal whorl and the stamens. Fruit a capsule or berry. Endosperm present; embryo small and straight.

Economic and ornamental plants. The family contains a number of ornamental plants. Many of these, e.g. *Narcissus* (Daffodil) and *Galanthus* (Snowdrop), are sufficiently hardy to be grown outside in the herbaceous border or rock-garden, while others, such as *Hippeastrum*, are popular house plants.

Classification. This family, distinguished by an inferior ovary, is treated here as separate from the Liliaceae, which has a superior ovary. Some authorities, however, regard the position of the ovary as less significant, and include the family within an enlarged Liliaceae. The genera listed here are grouped into 10 tribes.

A. Corona absent.
1. Galantheae (stem leafless; ovules many; perianth-tube absent or very short; flowers regular, solitary or few)
 Galanthus (12) Europe to Iran.
 Leucojum (9) Europe to Iran, Morocco.

2. Amaryllideae (stem leafless; ovules many; perianth-tube absent or very short; flowers more or less irregular, usually several)
 Nerine (30) S. Africa.
 Brunsvigia (20) S. Africa.
 Amaryllis (1) S. Africa.

3. Crineae (stem leafless; ovules many; perianth-tube present; flowers several)
 Crinum (130) tropics and subtropics.
 Cyrtanthus (47) tropical and S. Africa.

4. Zephyrantheae (stem leafless; ovules many; perianth-tube present; flowers solitary or paired)
 Zephyranthes (71) warm and tropical America, W. Indies.
 Sternbergia (7) S. Europe to Kashmir.

5. Haemantheae (stem leafless; ovules few)
 Haemanthus (21) S. Africa.
 Clivia (4) S. Africa.

6. Ixiolirieae (stem leafy below; umbel subcompound)
 Ixiolirion (4) S.W. and central Asia.

B. Corona present.
7. Euchariteae (corona of expanded, often united filaments)
 Hymenocallis (40) warm regions of America.
 Pancratium (16) Mediterranean region to tropical Asia and tropical Africa.
 Eucharis (10) tropical S. America.

8. Eustephieae (corona of small teeth present between filaments; perianth-lobes not spreading)
 Phaedranassa (6) Andes.

9. Hippeastreae (corona of scales; perianth-lobes spreading)
 Hippeastrum (75) tropical and subtropical America.
 Sprekelia (1) Mexico.

10. Narcisseae (corona distinct from filaments, annular or tubular or of separate scales)
 Narcissus (50) Europe, Mediterranean region, W. Asia.

177

23.1 Galanthus nivalis L. Snowdrop

Distribution. A native of damp woods and stream-sides from Europe to Iran. It is commonly cultivated for ornament, and widely naturalised.

Vegetative characteristics. A bulbous plant with the scape arising from between a pair of narrow, glaucous green leaves.

Floral formula. P3+3 A3+3 $\bar{G}(3)$

Flower and inflorescence. The solitary, pendulous, white flowers, with green horseshoe-shaped markings on their inner perianth-segments, appear from January to March. The fruit becomes fully developed in June.

Pollination. The family is well adapted to insect pollination owing to the conspicuous, often scented, flowers, and nectar-secreting glands. In *Galanthus*, the flower has 6 stamens which cluster round the longer style. At maturity the anther-cells split open at their tips, forming a hollow cone. When an insect, such as a bee, searches for nectar and touches this cone, it causes pollen to fall upon itself. In the course of foraging, the pollen is transferred to other flowers. In the absence of insect visits self-pollination is possible, as the cone of anthers becomes more lax, allowing the pollen to be shed directly on to the stigma below.

Alternative flowers for study. Any other species of *Galanthus* would be suitable, or any member of the closely related genus *Leucojum* (Snowflake) which is in flower between February and September according to species. Compare with genera in other tribes, e.g. *Nerine*, which flowers in the autumn.

Fig. 67. Flower of *Nerine bowdenii*.

A. The whole plant in flower. The ovoid bulb gives rise to a pair of glaucous green leaves enclosed below in a tubular. membranous sheath. Between the leaves is a long scape. The perianth-segments arise from the top of the prominent, green ovary. The outer, spreading segments are pure white. The inner segments are obovate, deeply notched, and have a green marking round the sinus.
Bulb: 15 x 10 mm Leaves: 115 x 4 mm
Scape: 130 mm Spathe: 27 mm

B. L.S. of flower. The flower arises from a spathe formed of 2 leaf-like, green bracteoles which are at first joined at their margins by 2 strips of membrane. One of these splits to allow the flower and pedicel to emerge, while the other shrinks in width and draws the 2 bracteoles together. The 6 perianth-segments are in 2 whorls. One segment from each whorl has been removed.
Perianth-segments (outer): 19 mm
 (inner): 11 mm
Ovary: 7-8 mm

C1. One of the 6 stamens, each of which is inserted opposite a perianth-segment. The anthers are 2-celled and face inwards. In most genera of the family they dehisce lengthwise, but in *Galanthus* and *Leucojum* the anthers split open at the apex forming a hollow cone.

C2. Dorsal view of anther.
Anther: 6.25 mm Filament: 2 mm

D. L.S. of the inferior ovary containing numerous ovules. The single style ends in a capitate stigma.

E. T.S. of the 3-locular ovary, showing some of the ovules attached to the axile placenta.
Ovary: 7 x 3.25 mm

F. The developing fruit, a capsule, bearing the persistent style. The seeds have a horn-like elaiosome which attracts ants, causing these insects to help in the dispersal of the seeds.
Fruit: 10 x 5 mm Style: *c*. 8 mm

Inner perianth-segment ('standard')

Branch of petaloid style

Bifid tip of style ('crest')

Stigma

Anther

Outer perianth-segment ('fall')

Perianth-tube

Stylar column

Inferior ovary

Spathe

Scape

Fig. 68. L.S. of *Iris* flower.

Flower of a bearded *Iris*

Iris pseudacorus, Yellow Iris

Inflorescence of a *Freesia*

Flower of *Sisyrinchium bermudiana*, Blue-eyed Grass

Fig. 69. Flowers of various genera of the Iridaceae.

24 IRIDACEAE Juss. Iris family

92 genera and 1850 species

Distribution. World-wide, but especially S. Africa, eastern Mediterranean region, Central and S. America.

General characteristics. Chiefly perennial herbs, with roots produced from a rhizome, bulb or corm. Leaves usually linear to sword-shaped, often overlapping in 2 ranks, with parallel veins. Inflorescence usually cymose. Flowers bisexual, regular or irregular, with spathes at their base. Perianth-segments 3 + 3, free or united, petaloid. Stamens 3 (representing the remaining outer whorl), with 2-celled anthers that face outwards and dehisce lengthwise. Ovary inferior, of 3 united carpels, usually 3-locular, with few to many anatropous ovules on axile placentas (unilocular with parietal placentation in *Hermodactylus*); style usually 3-branched, and frequently more or less petaloid. Fruit a capsule dehiscing by 3 valves; seed with copious, hard endosperm; embryo small.

Economic and ornamental plants. *Crocus sativus* is grown commercially, largely in Spain, for its stigmas which are the source of saffron, and the rhizome of *Iris germanica* var. *florentina* (Orris Root) is used in perfumery. The family contains numerous decorative plants, many of which are a prominent feature of gardens and parks. The genus *Iris* is particularly well represented by species, hybrids and cultivars, the group known as 'Bearded Irises' providing the main display. Other genera of horticultural value include all those mentioned in the section on Classification below.

Classification. The family is classified by some authorities into a considerable number of tribes, but based on the classification by Goldblatt (see References), it is divided here into 3 only:

1. Sisyrinchieae (stamens alternate with style branches; style with entire branches or unbranched; flowers enclosed in large bracts, few or several variously grouped together; rootstock not a corm or bulb)
 Sisyrinchium (100) Mostly America.
 Libertia (11) Australasia, Andes.

2. Irideae (stamens opposite style-branches; style usually branched, the branches divided; flowers enclosed in large bracts and variously grouped together, or sometimes solitary; rootstock a bulb, corm or rhizome)
 Iris (300) N. temperate region.
 Moraea (111) tropical and S. Africa.
 Tigridia (27) Mexico to Chile.

3. Ixieae (style always branched; inflorescence a spike of several flowers, or flowers solitary; rootstock usually a corm, rarely a rhizome)
 Gladiolus (180) W. and central Europe, Mediterranean region to central Asia, tropical and S. Africa.
 Romulea (95) Europe, Mediterranean region, S. Africa.
 Crocus (80) Europe to W. China.
 Ixia (45) tropical and S. Africa.
 Tritonia (28) tropical and S. Africa.
 Dierama (20) tropical and S. Africa.
 Freesia (11) S. Africa.
 Crocosmia (11) S. Africa.
 Sparaxis (6) S. Africa.

Fig. 70
Inflorescence (panicle) of
Dierama pulcherrimum,
Wand Flower.

24.1 Iris pseudacorus L. Yellow Flag

Distribution. Native in W. Asia, N. Africa and most of Europe including the British Isles where it is commonly found in marshes, swamps, ditches and at the edges of rivers and ponds. It has been introduced into N. America and has become naturalised in the central and eastern United States and eastern Canada.

Vegetative characteristics. An erect perennial, reaching a height of 40-150 cm, with a well-developed, fleshy rhizome. The sword-shaped leaves are about as tall as the scape.

Floral formula. P3+3 A3 \overline{G}(3)

Flower and inflorescence. Groups of 2 or 3 large, showy, yellow flowers are borne at the top of a tall, stout scape from May until July. Below each group is a green spathe with a papery margin.

Pollination. Bumble-bees and long-tongued flies are attracted to the flowers by the bright yellow perianth-segments and by the nectar that is secreted at their base. Functionally, the *Iris* flower is composed of 3 separate units. On visiting a flower, the insect lands on one of the outer perianth-segments ('falls'), which is marked with nectar-guides in the form of converging lines. As it moves forward, the insect bends back the flap-like stigma, transferring pollen which it has brought from another flower. Further on, it brushes against the anther, which dusts its body with fresh pollen. After it has extracted nectar, the insect creeps out backwards. Self-pollination cannot normally occur since the insect's body only touches the non-receptive side of the stigma as it leaves the flower. Insects have sometimes been seen to creep in or out between the style-branch and the fall. In this way they may obtain nectar without pollinating the flower.

Alternative flowers for study. Numerous species and hybrids of *Iris* are in cultivation and would be suitable alternatives. The commonly grown 'Bearded Irises' are so called because of the line of dense hairs at the base of the falls. In *Gladiolus* and *Freesia* the flowers are irregular.

A. The regular, bisexual flower. The yellow perianth is made up of 3 outer segments and 3 inner, united into a short tube above the inferior ovary. The outer segments ('falls') are narrow at their base but expand into a broad blade which hangs down. The inner segments ('standards') are more or less erect and smaller, but also have a well-defined limb and claw. Arising from the centre are the 3 petaloid style-branches which, for clarity, have been separated from the outer perianth-segments though in nature they lie close above them.
Perianth-segment (outer): 55 x 35 mm
 (inner): 25 x 5 mm
Petaloid style-branch: 40 x 5 mm

B1. L.S. of part of a flower with the perianth-segments separated to show the stamens inserted at the base of the outer segments. Two of the 3 inner segments are shown joined near the base with the outer segments. The inferior ovary, which contains numerous ovules attached to axile placentas, supports a column which divides into 3 petaloid style-branches and protects the stamens from above.
Ovary: 20-30 x 6-7 mm

B2. The tip of a style-branch, showing the non-receptive side of the stigma, a small flap situated on the underside, just below the divided tip ('crest'). The receptive surface appears when the flap is bent back.
Apex of style: 7 mm wide Lobe: 5 x 2.5 mm
Stigma: 1.5 x 1.5-1.75 mm

C. A stamen, situated opposite one of the perianth-segments and flanked by 2 of the inner perianth-segments.

D. Detail of anther which is 2-celled, faces outwards, and dehisces lengthwise.
Filament: 10 mm Anther: 15 mm

E. T.S. of ovary, showing the 3 loculi and the ovules attached to axile placentas.

F. The mature fruit, a capsule which dehisces by 3 valves to liberate the brown, flattened seeds. The seeds are buoyant and readily dispersed by water.
Fruit (before dehiscence): 45-50 x 15 mm
Seed: 10 x 8 mm

24.2 Crocus tommasinianus Herbert

Distribution. The species is native in parts of the Balkans. It grows in deciduous woods, on shady banks, and among limestone rocks at higher altitudes.

Vegetative characteristics. Readily reproducing by seeds as well as by corms. Each corm gives rise to 3 or 4 very narrow leaves with a distinctive silvery white line along the midrib.

Floral formula. P3+3 A3 \overline{G}(3)

Flower and inflorescence. The flowers appear from March to May in the wild but may be earlier in cultivation. The perianth-segments are pale lilac to purple, often silvery or buff on the outside, and are narrowed at the base into a long, white perianth-tube which serves as a stem. Variants include reddish purple forms, a lavender-coloured form with white tips ('Pictus'), and a white form ('Albus').

A. The whole plant at flowering stage. The corm is covered by a fibrous tunic. Surrounding the scape are several white membranous sheaths. The lowest is called the prophyll or basal spathe, and above this are 3 or 4 cataphylls. At the top is the greyish- or greenish-veined bract or floral spathe. These structures enclose the aerial shoot in the early stages.
Whole plant: 14.8 cm Corm: 12 x 15 mm
Scape: 25 mm Perianth-tube: 7.5 cm

B. Section of the upper part of a plant, showing the regular flower. The perianth consists of a long tube that opens out into 6 lobes, in 2 whorls of 3, the inner segments sometimes slightly broader than the outer. Part of the bract can be seen enveloping the narrow perianth-tube. The long 3-branched style is at an early stage of development.
Perianth-segments (outer): 41 x 15 mm
(inner): 41 x 17 mm

C. Lower part of the perianth-tube, showing the inferior ovary supported on a short subterranean scape.

D. The junction of a perianth-segment with the tube, showing the attachment of one of the 3 stamens. The anthers are 2-celled, dehisce lengthwise, and are attached at the base to the white filaments. The ring of hairs helps to protect the nectar from unwanted insects.
Filaments: 6.5-7 mm Anthers: 12-14 mm

Pollination. The flowers are closed at night and in dull weather, but in warm sun they open wide to reveal the orange style-branches and the yellow anthers. Long-tongued bees and butterflies forage for nectar secreted by the ovary and collected within the long perianth-tube. The insects collect large quantities of pollen from the stamens and carry it to the style-branches of other flowers.

Alternative flowers for study. In the absence of *C. tommasinianus,* another common spring-flowering species such as *C. vernus* is recommended. There are also a number of autumn-flowering species, including *C. nudiflorus* and *C. speciosus.* Both of these are leafless at flowering time.

E. T.S. of anther taken from a young flower-bud, showing an anther-cell on each side of the well-developed connective.
T.S. of anther: 0.8 x 0.4 mm

F. L.S. of the inferior ovary containing many ovules attached to axile placentas. Above the ovary can be seen the base of the long style and the perianth-tube. The outermost structure is the prophyll which joins the short scape at the base of the ovary.
Ovary: 5 mm

G. The apex of the style divides into 3 branches. The style-branches become expanded, flattened and fringed in the stigmatic region.
Style-branches: up to 5 mm

H. T.S. of the ovary after fertilisation, showing the 3 loculi, each containing 2 rows of young seeds attached to an axile placenta.
T.S. of ovary: 2.5 mm

I. The fruit before dehiscence. As the ovary develops into a capsule, the scape lengthens, pushing it above the surface of the ground. The capsule dehisces by 3 valves to shed the reddish brown seeds. The seeds have a prominent caruncle, which is attractive to ants. These insects carry the seeds away from the parent plant and disperse them over a wide area.
Fruits: 11-20 x 5-8 mm

25.1 *Dactylorhiza fuchsii*, Common Spotted Orchid

25 ORCHIDACEAE Juss. Orchid family

795 genera and 17500 species

Distribution. World-wide.

General characteristics. Perennial herbs, generally terrestrial (particularly in temperate regions) or epiphytic (mainly in the tropics), occasionally saprophytic (e.g. *Neottia*). Mycorrhiza often present. Sometimes climbing (e.g. *Vanilla*). Terrestrial species often having tubers or rhizomes; epiphytic species often having stems swollen into pseudobulbs. Leaves simple, usually alternate, often distichous, sometimes reduced to scales. Inflorescence a raceme or flowers solitary. Flowers usually bisexual, irregular, and extraordinarily diverse in size, shape, and colour. Perianth typically of 6 segments in 2 whorls of 3, the outer whorl (sepals) often petaloid, the inner petaloid and with a highly modified posterior petal, the labellum, which is often projected basally into a spur. Although the labellum or lip is strictly the uppermost petal, in most orchids it appears on the lower side of the flower. This is caused either by the ovary twisting through 180 degrees, or, in many single-flowered orchids, by the pedicel bending back over the top of the stem. The effect of this action, called resupination, is to enable the labellum to act as a landing-place for pollinating insects. Stamens usually 1, more rarely 2 or 3. Pollen grains usually grouped into pollinia. Ovary inferior, of 3 united carpels, usually unilocular, with numerous ovules on 3 parietal placentas. Style, stigmas and stamens variously united into a single, highly complex structure, the column. Stigmas basically 3, but only 2 fertile, the third represented by a sterile outgrowth, the rostellum. Fruit a capsule, containing a large number of seeds. The seeds are very small and light and are well suited to wind distribution. They have a minute embryo and little or no endosperm.

Economic and ornamental plants. The climbing genus *Vanilla*, especially the Central American species *V. planifolia*, is of economic importance and is now cultivated in many tropical countries for its pods, which are the source of the flavouring used in confectionery. The starchy tubers of certain genera, e.g. *Orchis*, *Dactylorhiza*, and *Eulophia*, are dried to form salep, which is used in parts of Europe and Asia for culinary and medicinal purposes. Many other members of the Orchidaceae are of horticultural value, mainly the epiphytic tropical orchids with their large, brightly coloured flowers. Genera commonly grown in greenhouses include *Cattleya*, *Paphiopedilum*, *Oncidium*, *Cymbidium*, *Dendrobium*, *Miltonia*, *Odontoglossum*, *Phalaenopsis*, and *Epidendrum*.

Orchid genera can be crossed easily and many bigeneric and plurigeneric hybrids are in existence.

Classification. While the characters of the Orchidaceae separate them clearly from other families, subdivision of the family has proved difficult and many schemes have been devised. The following classification, based on that given in Mabberley (see References), divides the family into 3 subfamilies:

I. Apostasioideae (flowers only slightly irregular; labellum shallow; stamens 2 or 3; pollen in separate grains)
 Apostasia (11) tropical Asia to Australia.

II. Cypripedioideae (flowers irregular; labellum slipper-shaped; stamens 2, staminode usually shield-like; pollen in separate grains)
 Paphiopedilum (36) tropical Asia to Solomon Islands.
 Cypripedium (35) N. temperate region.

III. Orchidoideae (flowers irregular; stamen 1; pollen grains grouped into pollinia)
 Dendrobium (900) tropical Asia, Australasia, Polynesia.
 Epidendrum (500) tropical America.
 Oncidium (412) tropical and subtropical America, W. Indies.
 Angraecum (206) tropical and S. Africa, Madagascar.
 Odontoglossum (200) Mexico to tropical S. America.
 Eulophia (200) mostly tropical Africa.
 Vanilla (100) tropics and subtropics.
 Cymbidium (40) tropical Asia to Australia.
 Phalaenopsis (40) Himalaya to Australia.
 Orchis (35) Madeira, temperate Eurasia to India and S.W. China.
 Vanda (35) China, Indomalaysia.
 Ophrys (30) Europe, W.Asia, N. Africa.
 Dactylorhiza (30) Atlantic Islands, N. Africa, temperate Eurasia, Alaska.
 Cattleya (30) Central and S. America, W.Indies.
 Miltonia (20) tropical America.
 Neottia (9) temperate Eurasia.

25.1 Dactylorhiza fuchsii (Druce) Soó Common Spotted Orchid

Distribution. Native throughout most of N. and central Europe and N. Asia. It is found in a wide range of habitats including grassy slopes, damp meadows, marshes, fens, and open woods.

Vegetative characteristics. A perennial herb with a palmately divided root-tuber. The leaves are keeled and sometimes slightly folded, obovate-oblong to elliptical, occasionally unmarked but usually heavily marked with elongated dark blotches.

Floral formula. K3 C3 A1 \overline{G}(3)

Flower and inflorescence. The inflorescence consists of a dense, many-flowered raceme. The flowers, which appear from May until early August, range from pale pink to mauve or white in colour, with red or purplish dots or lines on the lip.

Pollination. Many kinds of insects are attracted by the brightly coloured flowers. A visiting insect alights on the large horizontal petal (lip or labellum) which is prolonged backwards into a narrow spur. As the insect inserts its proboscis into the spur to obtain the liquid within its fleshy walls, it touches the bursicle, the pouch-like base of the rostellum. This ruptures, exposing 2 sticky discs (viscidia) one or both of which will adhere to the head of the insect. Each viscidium is attached to a group of pollen grains (pollinium) by threads collectively termed the caudicle. As the insect flies away, the membrane forming the top of the viscidium dries out, causing each pollinium to be directed forwards. By the time the insect has arrived at another flower, the pollinia are in the correct position to be transferred to the 2 sticky stigmas situated at the entrance of the spur.

Alternative flowers for study. Any other species of *Dactylorhiza* would be suitable, but in the interests of conservation it is important to check that the species chosen (even *D. fuchsii*) is locally common and that the minimum number of flowers is collected. Students in N. America may find species of *Corallorhiza* or *Spiranthes* more easily obtainable.

A. The inflorescence, a dense raceme of flowers which arise from the axils of leaf-like bracts.
Inflorescence: 5 cm

B. A flower. The 3 sepals and 3 petals arise from the top of the ovary. One of the petals is modified to form the 3-lobed labellum. The other petals overlap slightly and help to protect the reproductive organs which combine to form a structure known as the column.
Median or dorsal sepal: 7 x 2.5 mm
Lateral sepals: 7-8 x 3 mm
Labellum: 7-9 x 9-10 mm
Lateral petals: 5 x 2 mm

C. Side view of a flower with part of the perianth removed. During growth the ovary twists in a half-circle (shown by the ribs on its surface) bringing the labellum into a suitable position for insect visitors. The labellum is prolonged backwards into a spur. One of the 2 lateral petals can be seen behind the column, and arching over these parts is the median sepal.
Spur: 4-6 mm Ovary: 8 mm

D. The upper portion of the column, showing the stamen. The pollen masses (pollinia) are enclosed in membranes (thecae) situated at each side of the large, arching connective. At maturity, each theca splits open to expose the pollinium. At the base of the stamen is the rostellum which is prolonged upwards between the 2 thecae. Each theca narrows towards its base into a pouch-like flap, the bursicle, which protects from air the sticky disc (viscidium). The concave areas to the left and right of the pair of viscidia are the 2 stigmas and just above these are the 2 staminodes.
Stamen: 2 x 1 mm

E. The pollinium, a club-shaped mass of pollen grains joined together by elastic threads. The lower portions of the threads form a stalk, the caudicle, which connects the pollinium to the viscidium.
Pollinium: 1.5 mm

F. T.S. of the inferior, unilocular ovary with numerous ovules on the 3 parietal placentas.
Ovary: 1.25 mm in diameter

G. L.S. of the ovary, showing 2 of the 3 placentas bearing ovules. The ovary develops into a capsule which at maturity splits into 6 parts. Three of these bear the placentas with the seeds attached, the others are merely strips of pericarp. All 6 remain joined at their tips as well as at the base. In dry weather, the capsule contracts, causing the slits to open and allowing the minute seeds to escape and be dispersed by the wind. But in damp weather the dry pericarp absorbs moisture and lengthens, closing the slits and preventing seed-dispersal.
Ovary (at late flowering stage): 11 x 4 mm

A

B

E

D

F

C

G

L.S. of *Dactylorhiza* flower with lateral petals removed

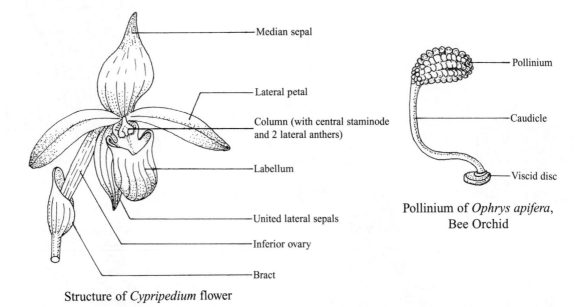

Structure of *Cypripedium* flower

Pollinium of *Ophrys apifera*,
Bee Orchid

Reproductive organs of *Cypripedium*

Pollen tetrad of *Neottia nidus-avis*,
Bird's-nest Orchid

Fig. 71. Structure of orchid flowers.

Aceras anthropophorum,
Man orchid

Anacamptis pyramidalis,
Pyramidal Orchid

Cephalanthera rubra,
Red Helleborine

Orchis simia,
Monkey Orchid

Goodyera repens,
Creeping Lady's-tresses

Gymnadenia conopsea,
Fragrant Orchid

Serapias vomeracea,
Long-lipped Serapias

Himantoglossum hircinum,
Lizard Orchid

Listera ovata,
Common Twayblade

Ophrys sphegodes,
Early Spider-orchid

Platanthera chlorantha,
Greater Butterfly-orchid

Spiranthes spiralis,
Autumn Lady's-tresses

Fig. 72. Flowers of various European orchids.

Fig. 73. Flowers of various cultivated orchids.

Comparative tables

Each of the following tables contains 3 families which have somewhat similar characters and which therefore may be confused with each other. Selected details of flowers, fruits, and in some cases vegetative parts are shown side by side in order to provide rapidly accessible information which should help the reader to distinguish more easily the families concerned.

Table 1

	1. Ranunculaceae	* Saxifragaceae	8. Rosaceae
Flower	Usually regular, rarely irregular (tribe Delphinieae); insect-pollinated	Usually regular, very rarely irregular (*Saxifraga stolonifera*); insect-pollinated	Regular, usually insect-pollinated
Calyx	Sepals variable in number, often petal-like (*Caltha*)	Sepals 4 or 5	Usually 5 free or united sepals, sometimes 4; epicalyx often present
Corolla	Petals variable in number, often absent (*Caltha*)	Usually 4 or 5 petals, sometimes united or absent	Usually 5 petals, absent in *Alchemilla* and *Sanguisorba*
Stamens	Numerous, arranged spirally or in whorls	Usually twice as many as petals, in 2 whorls, the inner whorl opposite the sepals, the outer whorl opposite the petals	Usually 2, 3, or 4 times as many as sepals and whorled; sometimes only 1-5 or numerous
Ovary	Superior, carpels 1 to many, usually free; 1 basal ovule or several marginal ovules	Superior to inferior; carpels usually 2, united; placentation axile with several rows of carpels	Usually superior, sometimes basally inferior (*Malus*); carpels 1 to many, usually free; placentation various
Fruit	Group of follicles or achenes, rarely a capsule (*Nigella*) or berry (*Actaea*)	Capsule or berry	Various: achenes (*Fragaria*) drupe (*Prunus*) follicle (*Spiraea*) pome (*Malus*)

Table 2

	14. Boraginaceae	15. Labiatae	16. Scrophulariaceae
Flower	Usually regular, rarely irregular (*Echium*)	Irregular	Irregular, sometimes nearly regular (*Verbascum*)
Calyx	5 united sepals, sometimes only basally united (*Lithospermum*)	5 united sepals, sometimes 2-lipped	Usually 5 united sepals (4 in *Veronica*)
Corolla	5 united petals	5 united petals, usually 2-lipped	Usually 5 united petals (4 in *Veronica*), often 2-lipped
Stamens	5	Usually 4 (one pair longer than the other pair), sometimes 2	Usually 4 (one pair longer than the other pair), sometimes 2, rarely 5 (*Verbascum*)
Ovary	2 united carpels becoming 4 at maturity	2 united carpels becoming 4 at maturity	2 united carpels
Fruit	Usually 4 nutlets separating at maturity, sometimes a drupe	Usually 4 nutlets separating at maturity, sometimes a drupe	Usually a capsule, sometimes a berry or drupe

Table 3

	20. Gramineae	*** Juncaceae**	*** Cyperaceae**
Inflorescence	Spike, raceme or panicle	Panicle or head	Various, from a simple spike to a much-branched panicle
Perianth-segments	Usually 2 or 3 scales (lodicules)	6 (in 2 whorls of 3), usually sepal-like	Often 6 bristles or scales, or absent
Stamens	Usually 3	Usually 6 (in 2 whorls of 3), or the inner whorl absent	Usually 1-3
Style(s)	Usually 2, 1 in *Nardus*, 3 in some bamboos	1, with 3 stigmas	1, with 2 or 3 stigmas
Fruit	Usually a caryopsis, rarely a nut or berry	Capsule	Achene, convex on both sides or triangular
Seeds	1	3 to many	1
Stem	Jointed, usually cylindrical, often with pith only at the nodes	Usually without joints, cylindrical, pith often present but not solid	Not jointed, usually 3-angled and with solid pith
Leaves	Basal (or if cauline 2-ranked), usually linear to lanceolate, and flat; sheath usually open	Mostly in basal tufts, usually linear, grass-like and flat, or cylindrical; sheath open or closed	In basal tufts (or if cauline 3-ranked), usually grass-like; sheath usually closed

Since these 3 families are sometimes particularly difficult to separate, some of their distinguishing characters have been illustrated in Fig. 63 at the end of the family Gramineae.

Table 4

	22. Liliaceae	23. Amaryllidaceae	24. Iridaceae
Inflorescence	Usually racemose, sometimes an umbel (*Allium*) or flowers solitary (*Tulipa*)	Cymose, often an umbel, sometimes flowers solitary (*Galanthus*)	Cymose, flowers sometimes solitary
Flower	Regular	Regular or irregular	Regular or irregular
Perianth-segments	3 + 3, petal-like or sometimes sepal-like	3 + 3, petal-like	3 + 3, petal-like
Stamens	Usually 3 + 3, anthers facing inwards	3 + 3, sometimes some sterile (staminodes), anthers usually facing inwards	3 (inner whorl absent), anthers facing outwards
Ovary	Superior; placentation usually axile	Inferior; placentation axile	Inferior; placentation usually axile
Fruit	Capsule or berry	Usually a capsule, sometimes a berry (*Clivia*)	Capsule
Storage organ	Rhizome, bulb, or fleshy root	Bulb or rhizome	Rhizome, bulb, or corm

Selected bibliography

Briggs, D. & Walters, S.M. *Plant Variation and Evolution*, 2nd edn, 1984. Cambridge University Press.

Davis, P.H. & Cullen, J. *The Identification of Flowering Plant Families*, 3rd edn, 1989. Cambridge University Press.

Engler, H.G.A. (1964). *Syllabus der Pflanzenfamilien*, vol.2, 12th edn, revised by H. Melchior. Gebrüder Borntraeger, Berlin.

Engler, H.G.A. & Prantl, K.A.E. (1st edn, 1887-1915; 2nd edn, 1924-). *Die natürlichen Pflanzenfamilien*. Leipzig and Berlin.

Goldblatt, P. (1971). *Journal of South African Botany* **37**, 371-460.

Heywood, V.H. (ed) (1978). *Flowering Plants of the World*. Oxford University Press.

Hickey, M. & King, C.J. *100 Families of Flowering Plants*, 2nd edn, 1988. Cambridge University Press.

Holmes, S. (1983). *Outline of Plant Classification*. Longman.

Holmes, S. (1987). *Handbook of Plant Types*. Hodder & Stoughton.

Hutchinson, J. (1923). *Kew Bulletin*, 65-89.

Ingrouille, M. (1992). *Diversity and Evolution of Land Plants*. Chapman & Hall.

Institute of Biology (1989). *Biological Nomenclature*. The Association of Science Education.

Jeffrey, C. (1982) *An Introduction to Plant Taxonomy*. Cambridge University Press.

Lawrence, G.H.M. (1962). *Taxonomy of Vascular Plants*. Macmillan Company, New York.

Mabberley, D.J. (1987). *The plant-book*. Cambridge University Press.

Mann, P. (1952). *Systematics of Flowering Plants*. Methuen & Co. Ltd.

Miller, R.N. (1989). *Plant Types 2: Mosses, Ferns, Conifers and Flowering Plants*. Hutchinson.

Proctor, M. & Yeo, P. (1973). *The Pollination of Flowers*. Collins (The New Naturalist Series).

Proctor, M., Yeo, P. & Lack, A. (1996). *The Natural History of Pollination*. Harper Collins.

Schumann, in Engler & Prantl.

Tutin, T.G. et al. (eds) *Flora Europaea*, vol.1 (1st edn, 1964; 2nd edn 1993). Cambridge University Press.

Weberling, F. (English translation by R.J. Pankhurst) (1989). *Morphology of Flowers and Inflorescences*. Cambridge University Press.

Wettstein, in Engler & Prantl.

Willis, J.C. (1973). *A Dictionary of the Flowering Plants and Ferns*, 8th edn, revised by H.K. Airy-Shaw. Cambridge University Press.

Wood, C.E. (1978). *A Student's Atlas of Flowering Plants: Some Dicotyledons of Eastern North America*. Harper & Row, U.S.A.

Zomlefer, W.B. (1994). *Guide to Flowering Plant Families*. University of North Carolina Press, Chapel Hill and London.

Glossary

Latin plurals: The following examples show how the plural is formed for most of the Latin terms in the glossary. Less common plurals, where required, are given after the terms concerned.

locul-us	locu-li
al-a	al-ae
haustori-um	haustori-a

abaxial The side of an organ away from the axis; dorsal.

abortion Non-formation or incompletion of a part.

abortive Imperfectly developed.

accrescent Increasing in size with age, as the calyx of some plants after flowering.

achene A small, dry, one-seeded, indehiscent fruit, strictly of one carpel, as in *Ranunculus* and *Rosa* (Fig. 25).

acicular Needle-shaped (Fig. 26).

actinomorphic Regular, divisible into equal halves in 2 or more planes.

acuminate Narrowing gradually to a point.

acute Sharply pointed.

acyclic Arranged spirally rather than in whorls.

adaxial The side of an organ towards the axis; ventral.

adnate United (with a different part).

adventitious bud A bud that arises from any part of a plant other than the axil of a leaf.

adventitious root A root that arises from any part of a plant other than the primary root system (Fig. 4, Fig. 7, Fig. 8 and Fig. 9).

aestivation The arrangement of the calyx or corolla in the bud (Fig. 13).

agamospermy The production of seeds by asexual means.

aggregate fruit A fruit formed by the joining of several carpels that were separate in the flower, as in the genus *Rubus*.

ala A wing or lateral petal in some of the Leguminosae.

alate Winged.

albumen Nutritive material stored within the seed.

albuminous Possessing albumen.

alternate (leaves) Placed singly at different heights on the stem or axis (Fig. 30).

amphitropous (ovule) Curved, so that both ends are brought near to each other (Fig. 19).

amplexicaul (leaves) Clasping the stem, but not completely encircling it (Fig. 29).

anatropous (ovule) With the body inverted so that it lies alongside the funicle (Fig. 19).

androecium The male sex organs (stamens) collectively.

androgynophore A stalk bearing both androecium and gynoecium, as in many *Passifloraceae.

androphore A stalk bearing the androecium, as in some *Tiliaceae.

anemophilous Depending on the wind to convey pollen for fertilisation.

angiosperm A plant having its seeds enclosed in an ovary.

annual A plant that completes its life-cycle within one year.

annular Ring-like.

annulus Ring.

anterior Front, away from the axis.

anther The part of the stamen that produces pollen (Fig. 15).

anther-cell (theca) One of the pollen-sacs that comprise the anther; at first there are usually 4 (2 in each lobe) but before anthesis the tissue separating each pair disintegrates and the anther then appears 2-celled (Fig. 15).

anthesis Flowering.

antipetalous (stamens) The same number as, and opposite to, the corolla-segments (Fig. 14).

antisepalous (stamens) The same number as, and opposite to, the calyx-segments (Fig. 14).

apetalous Without petals.

apex (plur. **apices**) The tip of an organ; adj., **apical**.

apiculus A short, sharp point.

apocarpous Having free carpels (Fig. 18).

apomixis Reproduction, including vegetative propagation, which does not involve sexual processes, often used in the narrower sense of agamospermy; adj., **apomictic**.

apopetalous With free petals.

aposepalous With free sepals.

appendage An attached subsidiary part.

arborescent Tree-like in growth or general appearance.

areole One of the small, spine-bearing areas on the stem of a cactus.

aril An outgrowth of the funicle, forming an appendage or outer covering on a seed.

arillate With an aril.

arilloid Resembling an aril.

arista An awn or stiff bristle.

aristate With an arista.

articulated Jointed.

ascending Directed upwards at an oblique angle.

asepalous Without sepals.

asexual Non-sexual.

atypical Not conforming to type.

auricle A small lobe or ear-shaped appendage; adj., **auriculate**.

awn A bristle-like appendage, often occurring on the glumes or lemmas of grasses; the strip of tissue attached to a mericarp in Geraniaceae.

axil The angle formed by the upper side of a leaf and the stem.

axile placentation A type of placentation in which the ovules are borne on placentas on the central axis of an ovary having 2 or more loculi (Fig. 20).

axillary In the axil.

axis (plur. **axes**) An imaginary line, round which the organs are developed.

baccate Berry-like.

basal At the base of an organ.

basifixed (anther) Attached by the base to the filament (Fig. 15).

bast Fibrous tissues serving for support.

beak A pointed projection.

berry A fleshy, indehiscent fruit with the seed or seeds immersed in pulp (Fig. 25, Fig. 42 and Fig. 57).

biconvex Convex on both sides, lenticular.

biennial A plant that completes its life-cycle in 2 years, flowering in its second year.

bifid Divided to about half-way into 2 parts.

bilabiate With 2 lips.

bilocular With 2 loculi.

bilobed With 2 lobes.

bipinnate Pinnate, with the primary leaflets again pinnate (Fig. 27).

biserrate Serrate, with the teeth themselves serrate.

bisexual Having both stamens and carpels in the same flower.

biternate Consisting of 3 parts, each part again divided into 3 (Fig. 27).

bivalved With 2 valves.

blade The expanded part of a leaf; the lamina.

bract A much-reduced leaf, especially the small or scale-like leaves associated with a flower or flower-cluster (Fig. 2).

bracteate Bearing bracts.

bracteole A small bract.

bulb A usually underground organ, consisting of a short, disc-like stem bearing fleshy scale-leaves and one or more buds, often enclosed in protective scales (Fig. 9); adj., **bulbous**.

bulbil A small bulb, often arising from the axil of a leaf or the inflorescence (Fig. 9).

bursicle The pouch-like base of the rostellum in some Orchidaceae.

caducous Falling off early.

caespitose Tufted.

calcicole Growing on soils containing lime.

calcifuge Growing on lime-free soils.

callus A hard or tough swollen area.

calyculus The epicalyx, appearing as a small rim or fringe in some of the *Loranthaceae.

calyx (plur. **calyces**) The outer perianth, composed of free or united sepals.

campanulate Bell-shaped (Fig. 12).

campylotropous (ovule) With the body curved so that it appears to be attached by its side to the funicle (Fig. 19).

capitate Pin-headed, as the stigma in the genus *Primula* (Fig. 18); growing in heads, as the flowers of Compositae.

capitulum A head of sessile or almost sessile flowers (Fig. 11); adj., **capitular**.

capsule A dry, dehiscent fruit formed from a syncarpous ovary (Fig. 25); adj., **capsular**.

carina A keel, as that formed by the 2 united lower petals in some of the Leguminosae.

carpel The unit of which the gynoecium is composed; in an apocarpous gynoecium the individual carpels are separate, in a syncarpous one the carpels are joined in various ways (Fig. 18).

carpophore The stalk to which the carpels are attached, as in Geraniaceae and Umbelliferae.

cartilaginous Gristly.

caruncle A protuberance near the hilum of a seed (Fig. 21); adj., **carunculate**.

carunculoid Resembling a caruncle.

caryopsis (plur. **caryopses**) The fruit of grasses, a one-seeded, indehiscent fruit with the pericarp united to the seed (Fig. 25).

cataphyll One of the papery, sheathing leaves which enclose the whole of the newly developing aerial shoot in the genus *Crocus* (Fig. 9).

catkin A spicate, often pendulous inflorescence of unisexual, apetalous flowers (Fig. 38).

caudicle The group of threads connecting a pollinium with the rostellum (Fig. 71).

cauline Borne on the stem.

chalaza The basal portion of the nucellus of an ovule (Fig. 19); adj., **chalazal**.

chasmogamy The opening of the perianth at the time of flowering; adj., **chasmogamous**.

chromosomes Thread-like structures which normally occur in 2 sets (diploid) in the nucleus of all vegetative cells and carry the genes or units of heredity; the sexual reproductive cells usually contain one set (haploid).

ciliate Fringed with hairs.

circumscissile Dehiscing by a line round the fruit or anther, the top coming off as a lid.

cladode (phylloclade) A branch taking on the form and functions of a leaf, as in some Euphorbiaceae and *Ruscus* in Liliaceae (Fig. 32).

clavate Club-shaped, thickened towards the apex.

claw The narrowed base of some petals.

cleistogamy Self-fertilisation within the unopened flower, as in some *Violaceae; adj., **cleistogamous**.

clone A group of plants that have arisen by vegetative reproduction from a single parent and which are therefore genetically identical.

coalescence Growing together; adj., **coalescent**.

coleoptile The sheath that surrounds the plumule in grasses, and protects it as it grows to the surface of the soil (Fig. 21 and Fig. 24).

coleorhiza The sheath that surrounds the radicle in grasses (Fig. 21).

collateral Standing side by side.

columella The persistent central axis round which the carpels of some fruits are arranged.

column The combination of stamens and styles into a single structure, as in *Asclepiadaceae and Orchidaceae.

coma The tuft of hairs at the end of some seeds.

commissure The face by which 2 carpels adhere, as in Umbelliferae.

compound Composed of 2 or more similar parts.

concave Appearing as if hollowed out.

conduplicate Folded together lengthwise.

confluent Merging together.

conifer A plant bearing cones, as the Pine.

connate United (with a similar part) (Fig. 29).

connective The part of the stamen that joins the 2 pairs of anther-cells (Fig. 15).

connivent Converging.

contiguous Touching or adjoining.

contorted Twisted (Fig. 13).

contractile root A root that can shorten in order to keep a bulb, corm or rhizome at a particular level (Fig. 8 and Fig. 9).

convex Having a rounded surface.

convolute Rolled together lengthwise (Fig. 13).

cordate Heart-shaped (Fig. 26).

coriaceous Leathery.

corm A swollen underground stem, somewhat bulb-like in appearance, but solid and not composed of fleshy scale-leaves (Fig. 9)

corolla The inner perianth, composed of free or united petals.

corona A structure occurring between (and sometimes united with) the stamens and the corolla, as in *Asclepiadaceae and Amaryllidaceae.

corpusculum The clip connecting the 2 bands (retinacula) which are attached to the pollinia in *Asclepiadaceae.

cortex Root or stem tissue lying between the medulla and the outermost layer (Fig. 7).

corymb A racemose inflorescence with pedicels of different lengths, causing the flower-cluster to be flat-topped (Fig. 11); adj., **corymbose**.

cotyledon One of the first leaves of the embryo (Fig. 21, Fig. 22 and Fig. 23).

coumarin An aromatic substance, the smell of newly mown hay, especially of the grass *Anthoxanthum odoratum*.

crenate With rounded teeth (Fig. 28).

crest An elevation or ridge on the summit of an organ (Fig. 68).

cross-pollination The transfer of pollen from one plant to another.

cucullus A corona-hood in the *Asclepiadaceae.

culm The hollow stem of grasses.

cultivar A horticultural variety that has originated and persisted under cultivation, not necessarily referable to a botanical species.

cuneate Wedge-shaped.

cupule A cup-shaped structure composed of coalescent bracts, as in the *Fagaceae; adj., **cupular**.

cuticle The waxy covering secreted by the cells of the epidermis.

cyathium The inflorescence of the genus *Euphorbia*.

cyclic Arranged in whorls.

cyme A determinate inflorescence (Fig. 10); adj. **cymose**.

cymule A small cyme or portion of one.

cypsela A small, dry, one-seeded, indehiscent fruit formed from 2 united carpels, as in Compositae (Fig. 25).

cystolith A crystal or deposit, usually of calcium carbonate, within a cell.

deciduous (leaves) Falling in autumn; (other organs) falling before the majority of the associated organs.

decumbent Lying along the ground, but with the apex ascending.

decurrent Running down, as when the base of a leaf is prolonged down the stem as a wing (Fig. 29).

decussate In opposite pairs, each pair at right angles to the next (Fig. 30).

dehiscence The process of splitting open at maturity.

dehiscent Splitting open along definite lines.

deltoid Triangular (Fig. 26).

dentate Toothed (Fig. 28).

denticulate Minutely toothed.

determinate (inflorescence) One in which the terminal flower opens first and prevents further growth of the axis.

diadelphous With 2 groups of stamens, as in some Leguminosae (Fig. 14).

dichasial cyme (dichasium) A cyme with lateral branches on both sides of the main axis (Fig. 10).

dichotomous Forked with 2, usually equal, branches.

dicotyledon A plant having 2 seed-leaves; adj., **dicotyledonous**.

didynamous Having one pair of stamens longer than the other pair, as in most Labiatae (Fig. 14).

digitate Palmate with narrow leaflets.

dimerous (2-merous) Having the parts of the flowers in twos.

dimorphic Occurring in 2 forms.

dioecious With male and female flowers on separate plants.

diploid Having 2 sets of chromosomes in each cell.

diplostemonous Having the stamens in 2 whorls, the inner whorl opposite the petals and the outer whorl opposite the sepals.

disc A development of the receptacle within the perianth, often bearing nectar-glands; the central part of the capitulum in the Compositae.

disc-floret One of the central, regular flowers in the flower-heads of some Compositae.

discoid Resembling a disc.

distichous Arranged in 2 vertical ranks.

distinct Separate.

dithecous With 2 anther-cells.

divergent Spreading away from each other.

divided (leaves) With lobing extending to the midrib or the base.

dorsal The side of an organ away from the axis; abaxial.

dorsifixed (anther) Attached by the back to the filament (Fig. 15).

double (flowers) Having more than the usual number of petals, usually at the expense of the stamens.

drupaceous Resembling a drupe.

drupe A fleshy fruit containing one or more seeds, each enclosed within a stony endocarp, as in the genus *Prunus*, (Fig. 25).

drupelet One of the small drupes which together form an aggregate fruit, as in the genus *Rubus* (Fig. 25).

ebracteolate Without bracteoles.

eglandular Without glands.

elaiosome An appendage on the seeds of some plants (e.g. *Helleborus*, *Viola*) which contains oily substances attractive to ants.

ellipsoidal Elliptical in outline with a 3-dimensional body.

elliptic(al) In the form of an ellipse (Fig. 26).

emarginate Having a notch in the margin at the apex.

embryo The rudimentary plant within the seed (Fig. 19, Fig. 21 and Fig. 22).

endocarp The innermost layer of the pericarp.

endosperm The albumen of a seed, particularly that deposited within the embryo sac (Fig.21); adj., **endospermic**.

ensiform Sword-shaped (Fig. 26).

entire Without division or toothing of any kind (Fig. 28).

entomophilous Depending on insects to convey pollen for fertilisation.

epicalyx (plur. **epicalyces**) A whorl of sepal-like organs just below the true sepals.

epicarp The outermost layer of the pericarp.

epicotyl The part of a seedling above the cotyledon(s) which gives rise to the stem and leaves (Fig. 23).

epidermis The outermost cellular layer of a plant; adj., **epidermal**.

epigeal A type of germination in which the cotyledons emerge above the ground (Fig. 23).

epigynous With the sepals, petals and stamens inserted near the top of the ovary (Fig. 17).

epipetalous Borne upon the petals (Fig. 14).

epiphyte A plant which grows on another plant but does not derive nourishment from it; adj.,**epiphytic**.

equitant Overlapping in 2 ranks, as the leaves in the genus *Iris*.

ericoid Resembling *Erica*.

exalbuminous (seeds) Lacking endosperm.

excentric Not having its axis placed centrally.

exserted Protruding.

exstipulate Without stipules.

extrafloral Outside the flower, as nectaries situated on the petiole.

extrastaminal Situated outside the whorl of stamens.

extrorse (anthers) Facing and opening outwards, away from the centre of the flower (Fig. 16).

falcate Sickle-shaped (Fig. 26).

fall One of the 3 outer perianth-segments in *Iris*.

family A group of genera resembling one another in general appearance and technical characters.

fascicle A close cluster or bundle; adj., **fasciculate**.

fertile Able to reproduce sexually.

fertilisation The effect of pollen, deposited on the stigma or stigmatic surface, causing the change of flower into fruit and ovule into seed.

fibrous roots Thread-like roots, particularly the adventitious roots arising from the base of grass stems.

filament A thread, particularly the stalk of an anther (Fig. 15).

filiform Thread-like.

fimbriate Fringed, having the margin cut into long, slender lobes.

floral Relating to the flower.

floret A small flower as in, e.g., Compositae and Gramineae.

flower The structure in angiosperms concerned with sexual reproduction.

foliaceous Leaf-like.

follicle A dry fruit formed from a single carpel, containing more than one seed, and splitting open along the suture (Fig. 25).

free Not united with similar parts.

free-central placentation A type of placentation in which the ovules are borne on placentas on a free, central column within a unilocular ovary (Fig. 20).

fruit A matured ovary with its enclosed seeds and sometimes with attached external structures.

funicle The stalk connecting an ovule to its placenta (Fig. 19).

funnel-shaped (corolla) With the tube gradually widening upwards and continuing imperceptibly into the limb (Fig. 12).

fused Completely united.

fusiform Spindle-shaped.

fusion Complete union of parts.

gamopetalous With united petals (Fig. 12).

gamosepalous With united sepals.

genes Hereditary units carried in chromosomes.

genetics The study of variation and heredity.

geniculate Abruptly bent.

genus (plur. **genera**) A group consisting of one or more similar species.

girdle scar The scar left by the terminal bud of the previous year (Fig. 3).

glabrous Without hairs.

gland A secretory organ (Fig. 48).

glandular Possessing glands.

glaucous With a waxy, greyish blue bloom.

globose Spherical.

glochid A barbed hair or bristle.

glomerule A condensed cyme of almost sessile flowers; a compact cluster.

glume One of the pair of bracts at the base of a spikelet in Gramineae; the single bract subtending the flower in *Cyperaceae.

gymnosperm A plant having its seed naked, i.e., not enclosed in an ovary, as a conifer.

gynobasic (style) Arising from below the ovary and between the carpels, as in Boraginaceae and Labiatae.

gynoecium The female sex organs (carpels) collectively.

gynomonoecious Having female and bisexual flowers on the same plant.

gynophore The stalk bearing a carpel or gynoecium.

gynostegium The staminal crown in *Asclepiadaceae.

gynostemium The column formed by the androecium and gynoecium combined, as in *Aristolochiaceae.

half-inferior (ovary) Having the lower part embedded in the pedicel and the upper part exposed, as in many species of *Saxifraga*.

halophytic Growing on saline soils.

haploid Having a single set of chromosomes in each cell.

haplostemonous With a single series of stamens in one whorl.

hastate Arrowhead-shaped, but with the basal lobes directed outwards (Fig. 26).

haustorium A sucker of parasitic plants.

head A short, dense spike of flowers, the capitulum in Compositae.

herb A non-woody plant, or one that is woody only at the base; adj., **herbaceous**.

hermaphrodite Bisexual, having both stamens and carpels in the same flower.

heterogamous With each flower-head composed of 2 or more kinds of flowers, as in many species of Compositae.

heterostyly Variation in the length of the style (and stamens) in different flowers within a species, as in Primulaceae and *Lythraceae; adj., **heterostylous**.

hexamerous (**6-merous**) Having the parts of the flower in sixes.

hilum The scar left on a seed where it was previously attached to the funicle (Fig. 23).

hip The fruit of the Rose, fleshy, hollow, and enclosing achenes (Fig. 25).

hirsute Covered in rough, coarse hairs.

hispid Having stiff, bristly hairs.

homogamous With only one kind of flower; with anthers and stigmas maturing simultaneously.

honey-leaf A nectary, as in many Ranunculaceae.

hybrid A plant resulting from a cross between 2 or more genetically unlike plants.

hygrophilous Requiring abundant moisture.

hygroscopic Extending or shrinking according to changes in moisture content.

hymenopterous (insects) Having membranous wings.

hypanthium A cup-shaped or tubular enlargement of the receptacle or the bases of the floral parts (Fig. 46).

hypocotyl The part of a seedling below the cotyledons which gives rise to the root (Fig.8 and Fig. 23).

hypocrateriform Salver-shaped (Fig. 12).

hypogeal A type of germination in which the cotyledons remain underground (Fig. 23 and Fig. 24).

hypogynous With the sepals, petals and stamens attached to the receptacle or axis below the ovary (Fig. 17).

imbricate Overlapping (Fig. 13).

imbricate-ascending Having the standard petal inside the lateral petals (Fig. 13).

imparipinnate Pinnate, with a terminal leaflet (Fig. 27).

incised Cut deeply and sharply into narrow, angular divisions (Fig. 28).

indefinite (flower-parts) Of a large enough number to make a precise count difficult.

indehiscent (fruit) Remaining closed at maturity.

indeterminate (inflorescence) One in which the lower or outer flowers open first, and the axis continues growth.

induplicate Rolled or folded inwards.

inferior Below, as when the ovary appears embedded in the pedicel below the other floral parts (Fig. 5, Fig. 6 and Fig. 17).

infertile Not fertile.

inflexed Abruptly bent inward.

inflorescence The arrangement of the flowers on the floral axis; a flower-cluster.

infructescence A cluster of fruits, derived from an inflorescence.

inserted Growing out from another organ.

integument The outer covering of an ovule (usually 2 being present in angiosperms) which later hardens to become the testa, or seed-coat (Fig. 19).

internode The part of the stem between 2 nodes (Fig. 2 and Fig. 3).

intrastaminal Within the stamens.

introduced Brought in from another region.

introrse (anthers) Facing and opening inwards, towards the centre of the flower (Fig. 16).

involucel The involucre of a partial umbel in Umbelliferae.

involucre A ring of bracts surrounding the head of flowers in Compositae or subtending the umbel in Umbelliferae; adj., **involucral**.

irregular (flowers) Zygomorphic, divisible into equal halves in one plane only (Fig. 12, Fig. 41 and Fig. 56).

jaculator A hook-like outgrowth from the stalk of a seed which aids its dispersal, as in *Acanthaceae.

keel (**carina**) The 2 united lower petals in some flowers of the Leguminosae.

labellum A lip, especially the highly modified third petal in Orchidaceae.

labiate Lipped (Fig. 12), or a member of the Labiatae.

laciniate Cut into narrow lobes.

lamina Leaf-blade.

lanceolate Lance-shaped (Fig. 26).

lateral On or at the side.

latex Milky juice.

leaf A lateral outgrowth from the stem, usually consisting of a stalk (petiole) and a flattened blade (lamina).

leaflet A leaf-like segment of a compound leaf.

legume The 2-valved fruit formed from a single carpel in the Leguminosae (Fig. 25).

lemma The lower of the 2 bracts enclosing a grass flower.

lenticel A pore in the stem that allows gases to pass between the outside atmosphere and the interior of a plant (Fig. 3 and Fig. 8).

lenticular Lens-shaped, biconvex.

liane A woody, tropical climber.

ligulate Strap-shaped, or tongue-shaped (Fig. 12).

ligule A strap-shaped structure, such as the limb of the ray-florets in Compositae; the scarious projection from the top of the leaf-sheath in grasses (Fig. 29).

limb The expanded part of a gamopetalous corolla, as distinct from the tube.

limen The rim at the base of the androgynophore in *Passiflora*, Passion-flower.

linear Long and narrow, with parallel sides (Fig. 26).

lip One of the 2 divisions of a bilabiate corolla or calyx; the labellum in Orchidaceae.

lobe Any division of an organ, especially if the part is rounded.

lobed Having one or more lobes.

lobulate Having small lobes.

loculicidal (capsule) Dehiscing through the midrib of the carpels.

loculus Compartment of an ovary or anther.

lodicule One of usually 2 minute scales in a grass flower, generally considered to be a vestigial perianth.

lomentum A fruit derived from a single carpel which breaks-up into one-seeded portions, as in *Ornithopus* (Leguminosae) (Fig. 25).

lyrate Pinnatifid, with the terminal lobe rounded and much larger than the others (Fig. 26).

marginal placentation A type of placentation in which the ovules are borne on a placenta situated along the ventral suture of a single carpel (Fig. 20).

medulla The pith of a stem or central tissue of a root (Fig. 7).

membranous Thin and semi-transparent like a membrane.

mericarp A portion of a fruit which splits away as a perfect fruit, as in Geraniaceae and Umbelliferae.

mesocarp The middle layer of the pericarp.

micropyle The opening in the integuments of an ovule, through which the pollen-tube grows after pollination (Fig. 19 and Fig. 23); adj., **micropylar**.

microspecies Species founded on minute differences and used mostly for apomictic plants, as in the genera *Taraxacum* and *Hieracium*.

midrib The middle and principal vein of a leaf.

monadelphous Having the stamens united in one group by the fusion of their filaments, as in some of the Leguminosae, *Polygalaceae, and Malvaceae (Fig. 14).

moniliform Like a string of beads.

monochasial cyme (monochasium) A cyme with lateral branching on one side only of the main axis (Fig. 10).

monocotyledon A plant having one seed-leaf; adj., **monocotyledonous**.

monoecious With male and female flowers on the same plant.

monopodial With a simple main stem or axis, growing by apical extension and bearing lateral branches.

monothecous With one anther-cell.

monotypic Having only one representative, as a genus with one species.

mucilaginous Slimy.

mucronate Terminated abruptly by a short, straight point.

multicellular Many-celled.

multilocular With many loculi.

multiple fruit A fruit formed from an inflorescence, often including bracts, as in *Ananas* and *Morus* (Fig. 25).

mycorrhiza The association of fungi and the roots of plants to their mutual advantage.

myrmecophilous (plants) Providing food and sometimes shelter for ants and in return deriving benefit from their presence, as in some *Rubiaceae.

naked (flowers) Lacking a perianth.

naturalised Established thoroughly after introduction from another region.

nectar A sugar solution, attracting insects or birds to flowers for the purpose of pollination.

nectar-guide A line or other marking leading insects to the nectary.

nectariferous Producing nectar.

nectary A gland which produces nectar.

node The point on a stem where one or more leaves are borne (Fig. 2 and Fig. 3).

nucellus The mass of cells in the middle of the ovule which contain the embryo (Fig. 19).

nut A dry, one-seeded, indehiscent fruit with a woody pericarp (Fig. 25).

nutlet A small nut, sometimes applied to an achene or part of a schizocarp (Fig. 25).

obcordate Cordate, but having the notch at the apex, and the attachment at the point.

obdiplostemonous Having the stamens in 2 whorls, the inner whorl opposite the sepals and the outer whorl opposite the petals (Fig. 14).

oblanceolate Lanceolate, but attached at the narrower end (Fig. 26).

oblong Longer than broad, with nearly parallel sides (Fig. 26).

obovate Ovate, but attached at the narrower end (Fig. 26).

obtuse Blunt or rounded at the end.

ochrea A tubular sheath, formed by the fusion of 2 stipules at the nodes of many *Polygonaceae.

ochreate Having ochreae (Fig. 29).

octamerous (8-merous) Having the parts of the flower in eights.

offset A short runner producing a new plant at its tip, as in *Sempervivum*.

operculum The lid of a circumscissile fruit; the membranous cover of the nectar-secreting ring in *Passiflora*.

opposite (leaves) Two at a node, one on each side of the stem (Fig. 30).

orbicular Circular (Fig. 26).

ornithophilous Depending on birds to convey pollen for fertilisation.

orthotropous (ovule) Borne on a straight funicle (Fig. 19).

oval Broadly elliptical (Fig. 26).

ovary The lower part of a carpel (or carpels) which contains the ovules (Fig. 5 and Fig. 6).

ovate Shaped like the longitudinal section of a hen's egg (Fig. 26).

ovoid Egg-shaped.

ovule A structure in the ovary which, after fertilisation, develops into a seed (Fig. 19).

palate The swollen part of the lower lip of a gamopetalous corolla which almost or entirely closes the throat, as in *Antirrhinum*.

palea The upper of the 2 bracts enclosing a grass flower.

pali The shorter filaments in the corona of the *Passifloraceae.

palmate (leaves) Divided to the base into separate leaflets, all the leaflets arising from the apex of the petiole (Fig. 27).

pandurate Fiddle-shaped (Fig. 26).

panicle A much-branched inflorescence (Fig. 11); adj., **paniculate**.

papilla A minute protuberance.

papillose Bearing papillae.

pappus The specialised calyx of hairs or scales occurring mainly in the Compositae.

parasite A plant which lives on another plant and derives nourishment from it; adj., **parasitic**.

parenchyma Tissue composed of thin-walled cells with intercellular spaces; adj., **parenchymatous**.

parietal placentation A type of placentation in which the ovules are borne on placentas situated on the inner surface of the ovary-wall or on outgrowths from it (Fig. 20).

paripinnate Having an equal number of leaflets and lacking the terminal one (Fig. 27).

parted Cut, but not quite to the base.

parthenocarpy The formation of fruit without preliminary fertilisation and usually without development of seeds; adj., **parthenocarpic**.

patent Spreading.

pectinate Pinnatifid with narrow segments set close like the teeth of a comb.

pedicel The stalk of a single flower (Fig. 4, Fig. 5 and Fig. 6); adj., **pedicellate**.

peduncle The stalk of an inflorescence.

peltate (leaves) Having the blade attached by its lower surface to the stalk, instead of by its margin (Fig. 26).

pendent, pendulous Hanging down.

pentamerous (5-merous) Having the parts of the flower in fives.

pepo The unilocular, many-seeded fruit in *Cucurbitaceae (Fig. 25).

perennating Surviving from one season to another.

perennial A plant that lives for more than 2 years, flowering in its second and subsequent years.

perfoliate (leaves) Completely encircling the stem (Fig. 29).

perianth A collective term for the outer, non-reproductive parts of a flower, often differentiated into calyx and corolla.

pericarp The fruit-wall (Fig. 21 and Fig. 22).

perigynium The utricle which encloses the female flower in some *Cyperaceae.

perigynous With the sepals, petals and stamens inserted around the ovary on a concave structure developed from the receptacle (Fig. 17).

perisperm The albumen of a seed formed outside the embryo sac.

persistent Remaining attached.

personate (corolla) Bilabiate with a prominent palate (Fig. 12).

petal A single segment of the corolla (Fig. 5).

petaloid Petal-like.

petiolate Having a leaf-stalk.

petiole A leaf-stalk.

photosynthesis The process by which green plants convert carbon dioxide and water into carbohydrates in the presence of sunlight; adj., **photosynthetic**.

phylloclade (cladode) A branch taking on the form and functions of a leaf, as in some Euphorbiaceae and *Ruscus* in Liliaceae (Fig. 32).

phyllode A petiole taking on the form and functions of a leaf, as in *Acacia* (Fig. 32).

pilose Softly hairy.

pin-eyed A dimorphic flower, as *Primula vulgaris*, Primrose, where the style is long and the stamens are below the stigma.

pinna A primary division or leaflet of a pinnate leaf.

pinnate Having separate leaflets along each side of a common stalk (Fig. 27).

pinnatifid Pinnately divided, but not to the rachis.

pinnatisect Pinnately divided down to the rachis.

pistillate (flowers) Having only female organs.

placenta The part of the ovary to which the ovules are attached; adj., **placental**.

placentation The arrangement of placentas in an ovary (Fig. 20).

plantlet A small plant, as those formed on the margins of the leaves in some species of *Kalanchoe* (*Crassulaceae).

plica A fold.

plicate Folded, usually lengthwise.

plumose Feather-like.

plumule The part of the seed that develops into the shoot (Fig. 21 and Fig. 23).

pod A dry, many-seeded, dehiscent fruit, particularly the legume in Leguminosae.

pollen The small grains which contain the male reproductive cells of the flower.

pollination The placing of pollen on the stigma or stigmatic surface.

pollinium A pollen-mass, as in Orchidaceae and *Asclepiadaceae.

polygamo-monoecious Polygamous, but in the main monoecious.

polygamous Bearing both unisexual and bisexual flowers on the same plant.

polypetalous With a corolla of separate petals (Fig. 12).

polyploid Having more than 2 sets of chromosomes in each cell.

polysepalous With a calyx of separate sepals.

pome A fruit consisting of a core, formed by several united carpels, enclosed within the firm, fleshy receptacle, as in the genus *Malus*, Apple (Fig. 25).

pore A small aperture.

poricidal (anthers) Opening by pores, as in many of the Ericaceae and *Polygalaceae (Fig. 15).

posterior Back, towards the axis.

posticous On the posterior side, next the axis.

prickle A sharp-pointed outgrowth from the superficial tissues of the stem, as in the genus *Rosa* (Fig. 31).

proboscis The elongated part of the mouth of some insects.

procumbent Lying along the ground.

prophyll The lowest of the papery structures attached to the scape in the genus *Crocus* (Fig. 9).

prostrate Lying flat.

protandrous Having stamens which mature and shed their pollen before the stigmas of the same flower become receptive.

protogynous Having stigmas which become receptive before the stamens of the same flower mature and shed their pollen.

protoplasm The viscous, semi-transparent substance which forms the basis of all living plant-cells.

pseudobulb The thickened and bulb-like stem of some orchids.

pseudocarp A structure comprising the mature ovary combined with some other part of the plant, as the Rose 'hip' (Fig. 25).

pubescence Hairiness.

pubescent. Covered in soft hairs.

pulvinus An enlarged portion of the petiole, at its base in *Mimosa pudica* or at its junction with the lamina in *Marantaceae.

punctate Marked with dots, depressions, or translucent glands.

punctiform In the form of a point or dot.

pungent Ending in a stiff, sharp point.

pyxis (plur. **pyxides**) A capsule with circumscissile dehiscence, as in *Anagallis*.

quincuncial Partially imbricated of 5 parts, 2 being exterior, 2 interior, and the fifth having one margin exterior and the other interior (Fig. 13).

raceme An indeterminate inflorescence with pedicellate flowers (Fig. 10); adj., **racemose**.

rachilla A secondary axis in the inflorescence of grasses.

rachis (plur. **rachides**) The axis of a compound leaf or an inflorescence.

radical (leaves) Arising directly from the rootstock.

radicle The part pf the seed that develops into the root (Fig. 21, Fig. 23 and Fig. 24).

radii The longer filaments in the corona of the *Passifloraceae.

raphides Needle-shaped crystals found in the cells of some plants.

ray A primary branch of an umbel.

ray-floret One of the outer, irregular flowers in the flower-heads of some Compositae.

raylet A smaller, secondary branch of an umbel.

receptacle (**torus**) The end of the stem which bears the flower-parts (Fig. 5 and Fig. 6).

recurved Curved backward.

reflexed Bent backward abruptly.

regular (flowers) Actinomorphic, divisible into equal halves in 2 or more planes (Fig. 12).

reniform Kidney-shaped (Fig. 26).

replum The central partition in the fruits of Cruciferae.

resin A sticky, sometimes fragrant substance, insoluble in water, which is secreted by special cells in some plants for protective purposes; adj., **resinous**.

resupination Turning upside down, as flowers in many Orchidaceae; adj., **resupinate**.

reticulate Marked with a network pattern.

retinaculum The band connecting a pollinium to the corpusculum in *Asclepiadaceae.

revolute Rolled back from the margin or apex.

rhipidium A monochasial cyme, in which the lateral branches are developed alternately, resulting in a fan-shaped inflorescence.

rhizomatous Resembling or possessing a rhizome.

rhizome A root-like stem, lying horizontally on or situated under the ground, bearing buds or shoots and adventitious roots (Fig. 8).

rhomboid More or less diamond-shaped (Fig. 26).

rootstock A frequently subterranean stem or rhizome.

rosette A radiating cluster of leaves, usually at ground level.

rostellum The beak-like extension of the stigma in some Orchidaceae.

rostrum The beak-like, thickened base of the style in *Geranium*.

rotate (corolla) Gamopetalous, with a short tube and spreading lobes (Fig. 12).

rugose Wrinkled.

ruminate Looking as though chewed.

runcinate With sharply cut divisions directed backward towards the base (Fig. 26).

runner A creeping stem, rooting and giving rise to plantlets at the nodes.

sac A small pouch or bag-shaped structure; adj., **saccate**.

sagittate Arrowhead-shaped (Fig. 26).

salver-shaped (corolla) Having a slender tube expanding abruptly into a flat or saucer-shaped limb (Fig. 12).

samara A dry, indehiscent, winged fruit, as in *Fraxinus*, Ash (Fig. 25).

samaroid Resembling a samara.

sap The juice of a plant.

saprophyte A plant which lives upon dead organic matter; adj., **saprophytic**.

scabrid Rough.

scale A reduced leaf, usually membranous, and often found covering buds and bulbs.

scandent Climbing.

scape A leafless stalk, bearing one or more flowers, which arises from the ground.

scapose Having a scape.

scarious Thin, dry and membranous.

schizocarp A fruit derived from a syncarpous ovary which breaks up into one-seeded portions (mericarps) (Fig. 25); adj., **schizocarpic**.

scion A young shoot which is inserted into a rooted stock in grafting.

secund Directed towards one side.

seed A unit of sexual reproduction developed from the fertilised ovule.

segment One of the divisions of an organ.

sepal A single segment of the calyx (Fig. 5).

sepaloid Sepal-like.

septicidal (capsule) Dehiscing through the septa or carpel margins.

septifragal The kind of dehiscence where the valves break away from the partitions, as in *Impatiens* (*Balsaminaceae).

septum A partition; adj., **septate**.

serrate With a saw-toothed margin (Fig. 28).

serrulate Minutely serrate.

sessile Not stalked.

sheath A tubular covering.

shoot A developed bud or young green stem.

shrub A perennial, woody plant, generally smaller than a tree and having several main stems instead of a single trunk.

silicula The capsular fruit of the Cruciferae when less than 3 times as long as broad (Fig. 25).

siliqua The capsular fruit of the Cruciferae when at least 3 times as long as broad (Fig. 25).

simple Of one piece.

sinuate (leaves) Having the blade flat but with the margin winding strongly inward and outward (Fig. 28).

sinus The recess between 2 lobes, e.g., of a leaf.

solitary Borne singly.

spadix (plur. **spadices**) A spike with a fleshy axis, as in Araceae.

spathe A large bract subtending and often enclosing a flower or an inflorescence (Fig. 9, Fig. 64, Fig. 65 and Fig. 68).

spathulate Spatula-shaped (Fig. 26).

species (plur. **species**) A group of closely related, mutually fertile individuals, showing constant differences from allied groups, the basic unit of classification.

spicate Spike-like.

spike An indeterminate inflorescence with sessile flowers (Fig. 10).

spikelet A unit of the inflorescence in grasses, consisting of one or more flowers subtended by a common pair of glumes.

spine A sharp, woody or hardened outgrowth from a leaf, sometimes representing the entire leaf (Fig. 31).

spinose Spiny (Fig. 28).

spiralled cyme A monochasial cyme, in which the branching is always on the same side, resulting in a coiled inflorescence (Fig. 10).

spur A slender, hollow extension, often nectariferous, of the calyx or corolla (Fig. 35 and Fig. 71).

stamen One of the male sex organs, usually consisting of anther, connective and filament (Fig. 5 and Fig. 6); adj., **staminal**.

staminate (flowers) Having only male organs.

staminode A sterile stamen (Fig. 71); adj., **staminodial**.

standard The upper petal in some flowers of the Leguminosae, or one of the 3 inner perianth-segments in *Iris*.

stellate Star-shaped.

stem The main supporting axis of a plant.

sterile Unable to reproduce sexually.

stigma The apex of the style, usually enlarged, on which the pollen grains alight and germinate (Fig. 5 and Fig. 6); adj., **stigmatic**.

stipe The stalk of a stalked carpel or gynoecium.

stipulate Having stipules (Fig. 29).

stipule A leafy outgrowth, often paired, arising at the base of the petiole (Fig. 2, Fig. 31 and Fig. 32).

stock The rooted stem into which the scion is inserted when grafted.

stolon A lateral stem growing horizontally at ground level, rooting at the nodes and producing new plants from its buds, as in *Fragaria* × *ananassa*, Strawberry.

stoloniferous Bearing stolons.

stoma (plur. **stomata**) A breathing pore in the epidermis of a leaf or young stem.

stone The hard endocarp of a drupe.

style The often elongated apical part of a carpel or gynoecium bearing the stigma at its tip (Fig. 5, Fig. 6 and Fig. 18); adj., **stylar**.

stylopodium The enlargement at the base of the styles in Umbelliferae.

subapical Almost at the apex.

subbasal Almost at the base.

subcompound More or less compound.

subcordate More or less heart-shaped.

subcuneate More or less wedge-shaped.

subequal Almost equal.

subfamily A subdivision of a family, usually found desirable in large families with many components.

subgenus A subdivision of a genus.

subglabrous Almost without hairs.

subglobose Almost globular.

suborbicular Almost circular.

subsessile Almost devoid of a stalk.

subshrub A low shrub, sometimes with partially herbaceous stems; an undershrub.

subspecies A subdivision of a species, often used for a geographically or ecologically distinct group of plants.

subtend To stand below and close to, to extend under.

subulate Awl-shaped (Fig. 26).

subvalvate Almost valvate.

succulent Fleshy and juicy.

sucker A shoot of subterranean origin.

suffrutescent Woody only at the base of the stem.

suffruticose Woody in the lower part of the stem.

sulcate Grooved or furrowed.

superior Above, as when the ovary is situated above the other floral parts on the receptacle (Fig. 5, Fig. 6 and Fig. 17).

surmounted Capped or crowned.

suture A seam or line of joining.

syconium A multiple hollow fruit, as the Fig.

sympodial With the main stem or axis ceasing to elongate but growth being continued by the lateral branches.

synandrium An androecium coherent by the anthers, as in some Araceae.

syncarpous Having united carpels (Fig. 18).

syngenesious With anthers united into a tube, but filaments free, as in Compositae.

taproot A strongly developed main root which grows downwards bearing lateral roots much smaller than itself (Fig. 2, Fig. 7 and Fig. 8).

taxon (plur. **taxa**) A classificatory unit of any rank, e.g., Daisy, *Bellis perennis* (species); *Bellis* (genus); Compositae (family).

tendril A twining, thread-like structure produced from a stem or leaf which enables a plant to hold its position securely (Fig. 32).

tepal One of the petals or sepals of a flower in which all the perianth-segments closely resemble each other.

terete Slender and cylindrical, usually tapering, but more or less circular in any cross-section.

terminal At the apex or end.

ternate In threes.

testa The outer coat of the seed, formed by the hardened integuments (Fig. 21, Fig. 22 and Fig. 23).

tetrad (pollen) A group of 4 cells which remain together until they are almost or quite mature (Fig. 71).

tetradynamous Having 4 long stamens and 2 short, as in Cruciferae (Fig. 14).

tetramerous (4-merous) Having the parts of the flower in fours.

theca The pollen sac of an anther (Fig. 15).

thorn A short, pointed branch (Fig. 31).

throat The opening into a corolla or calyx which has united segments.

thrum-eyed A dimorphic flower, as *Primula vulgaris*, where the style is short and the stamens are above the stigma.

thyrse A mixed inflorescence in which the main axis is indeterminate and the secondary and ultimate axes are determinate or cymose.

tomentose Densely covered in soft hairs.

torus The receptacle of a flower.

translator The clip (corpusculum) and bands (retinacula) which connect the pair of pollinia in *Asclepiadaceae.

translucent Allowing light to pass through, but not transparent.

tree A perennial, woody plant with an evident trunk.

tribe A subdivision, usually of a large subfamily, but sometimes a family may be divided directly into tribes.

trichotomous Forked, with 3 usually equal branches.

trifid Divided to about half-way into 3 parts.

trifoliate Three-leaved.

trifoliolate With 3 leaflets (Fig. 27).

trigonous Three-angled.

trilobate (trilobed) With 3 lobes.

trimerous (3-merous) Having the parts of the flower in threes.

trimorphic Occurring in 3 forms.

triploid Having 3 sets of chromosomes in each cell.

truncate Appearing as if cut off at the base or apex.

tuber An underground stem or root, swollen with reserves of food, as *Solanum tuberosum*, Potato.

tubercle A small tuber; a small, rounded projection.

tubercled Warty.

tuberous Resembling or producing tubers.

tubular Cylindrical and hollow (Fig. 12).

umbel A usually flat-topped inflorescence in which the pedicels arise from the same point on the peduncle, as in most Umbelliferae (Fig. 11); adj., **umbellate**.

umbelliform In the shape of an umbel.

undershrub A low shrub, sometimes with partially herbaceous stems; a subshrub.

undulate (leaves) With a wavy margin curving up and down (Fig. 28).

unguiculate Narrowed at the base into a claw.

unicellular One-celled.

unilateral One-sided.

unilocular With a single loculus.

unisexual (flowers) Having only male or female organs.

united Joined.

urceolate Urn-shaped (Fig. 12).

utricle A bladder-like structure, especially the membranous sac in the fruit of *Carex*.

vaginate Sheathed (Fig. 29).

valvate (flower) When the parts of the corolla or calyx meet exactly without overlapping (Fig. 13).
(anther or fruit) Opening by valves (Fig. 15).

valve One of the pieces into which a capsule naturally separates at maturity.

variegated Having 2 or more colours in the leaves.

variety A rank subordinate to species used to designate a group of plants varying in flower-colour, habit, or some other way.

vascular bundle A strand of tissue which conducts water and nutrients (Fig. 2 and Fig. 4).

vein A strand of vascular tissue in a leaf or other flat organ.

venation The arrangement of veins, as in a leaf.

ventral The side of an organ towards the axis; adaxial.

verrucose Warty.

versatile (anther) Attached at the middle and turning freely on its filament (Fig. 15).

verticillaster A false whorl, as in Labiatae (Fig. 11).

verticillate Whorled.

vestigial Imperfect development of an organ which was fully developed in some ancestral form.

vexillary (aestivation) Having the vexillum folded over the other petals; imbricate-descending (Fig. 13).

vexillum The standard petal in the Leguminosae.

vicariad (vicariant) One of 2 similar taxa occupying separate geographical areas, e.g. *Jasminum nudiflorum* or *J. mesnyi* (*Oleaceae).

viscid (viscous) Sticky.

viscidium The sticky disc at the base of the caudicle in Orchidaceae.

viscin A sticky substance surrounding the seeds in *Loranthaceae.

vitta An oil-canal in the fruits of Umbelliferae.

whorl The arrangement of organs in a circle round the axis; adj., **whorled** (Fig. 30).

wing A lateral petal in some of the Leguminosae; a flat, often dry and membranous extension to an organ.

xerophyte A plant which can live in very dry conditions, as in a desert; adj., **xerophytic**.

zygomorphic Irregular, divisible into equal halves in one plane only.

Index of families and genera

Names of families treated in detail are in **bold** type; all other family names are in ordinary type.